"十四五"普通高等教育本科部委级规划教材
浙江省普通高校"十三五"新形态教材

Eco Technology of Leather Manufacture

皮革清洁生产技术

罗建勋 / 主编
马贺伟　彭必雨　[西]胡安·卡洛斯·卡斯特尔(Juan C. Castell) / 副主编

中国纺织出版社有限公司

内 容 提 要

本书着眼于皮革清洁生产技术,通过为轻化工程专业的学生讲授皮革工业发展史、皮革生产过程的知识及相关的清洁化技术,旨在使他们掌握这些知识和前沿技术,从而助推传统特色产业之皮革产业的改造提升和可持续发展。本书主要内容包括:皮革及其工业的发展史,制革用原料皮的种类、组织结构与化学组成,皮革之传统生产技术,皮革清洁化生产技术,皮革生产过程中产生固废的资源化利用技术,相关生态指标及标准等。

本书可作为高等院校轻化工程专业学生的教材,也可作为以皮革类研究为主的科研工作者和相关生产企业技术人员的参考书。

著作权合同登记号:01-2024-5782

图书在版编目(CIP)数据

皮革清洁生产技术 = Eco Technology of Leather Manufacture : 英文 / 罗建勋主编; 马贺伟, 彭必雨, (西) 胡安 · 卡洛斯 · 卡斯特尔副主编 . -- 北京: 中国纺织出版社有限公司, 2025. 1. -- ("十四五" 普通高等教育本科部委级规划教材) (浙江省普通高校 "十三五" 新形态教材) . -- ISBN 978-7-5229-2163-1

Ⅰ. TS5

中国国家版本馆 CIP 数据核字第 202415AA61 号

责任编辑: 孔会云　朱利锋　　责任校对: 高　涵
责任印制: 王艳丽

中国纺织出版社有限公司出版发行
地址: 北京市朝阳区百子湾东里 A407 号楼　邮政编码: 100124
销售电话: 010—67004422　传真: 010—87155801
http://www.c-textilep.com
中国纺织出版社天猫旗舰店
官方微博 http://weibo.com/2119887771
三河市宏盛印务有限公司印刷　各地新华书店经销
2025 年 1 月第 1 版第 1 次印刷
开本: 787×1092　1/16　印张: 12.5
字数: 289 千字　定价: 68.00 元

前　言

动物皮是畜牧业重要的副产品之一，将动物皮制成皮革或裘革是人类对其最重要的利用方式之一，是人类利用自然资源、追求自然本性、发展畜牧业经济、促进乡村振兴的重要途径。

利用动物皮革或裘革的历史源远流长，可追溯到至今300万年前，因此，一部人类的发展历史也就包含着不断持续改进的利用动物皮的过程。随着人们对自然界各种材料的逐步认知及后期对新材料的设计、合成与应用，皮革及裘革的质量也在持续提升，不断满足各时期人们对不同类型革制品质量的需求。

随着经济全球化的不断深入推进和习近平主席向全球提出的“一带一路”倡议的全面实施，全球皮革产业链经济非常活跃，大大促进了全球皮革产业的良性发展。

为了进一步贯彻党的二十大精神中“教育、科技、人才一体化建设”重要论述，使技术人员、科研工作者、皮革产业的从业者、轻化工程专业皮革方向的大学生和研究生等了解全球皮革产业的发展趋势，更好地认知皮革发展的历史、原料皮的相关知识、传统皮革生产过程、清洁化皮革生产技术、皮革生产过程中固废处理及相关标准等内容，特编写并出版《Eco Technology of Leather Manufacture：皮革清洁生产技术》。

本书共6章，第1章至第4章由嘉兴大学罗建勋教授编写，第5章和第6章由嘉兴大学马贺伟副教授编写，全书由四川大学彭必雨教授、胡安·卡洛斯·卡斯特尔（Juan C. Castell）博士指导。另外，在本书编写过程中，四川大学何有书，TRUMPLER公司的杰弗里·格斯里·斯特罗恩（Jeffry Guthrie-strachan）、华金·林佐安·阿塞多（Joaquin Linzoain Acedo）、张建维、AJ、巴努（BAHNU）、陈飞、陈正栓等，STAHL公司的蔡纯煜（Chusoon）、唐佩龙、夏敏，嘉兴大学易玉丹等提供了部分资料并给予大力支持，在此一并表示感谢。

鉴于最新皮革生产技术的飞速发展、查阅资料不全及作者个人水平的局限性等原因，本书可能存在不当之处，敬请读者指正。

罗建勋

2022年12月于嘉兴大学

Preface

Animal hide or skin is one of the important by-products of animal husbandry. Making animal hide or skin into leather or double-face is one of the most important ways to utilize it for humans, which is an important way for humans to utilize natural resources, pursue natural nature, develop animal husbandry economy, and promote rural revitalization.

The history of making animal hide or skin into leather or double-face can be traced back to over three million years ago. Therefore, the history of the development of human being also includes the process of continuously improving the use of animal skins. With the gradual understanding of various materials in nature and the subsequent design, synthesis, and application of new materials, the quality of leather and double-face was continuously promoted, constantly meet the quality needs for leather products of people in different periods for different types.

With the continuously advancing in depth of economic globalization and the full implementation of the Belt and Road Initiative proposed by President Xi Jinping, chain economy of the global leather industry is very active, which greatly promote the continuous development of the global leather industry.

In order to carry out the important statements of integrated construction of education, technology, and talent included in Report of the 20th National Congress of the Communist Party of China, cause the technologists, researchers, practitioners in the leather industry, university students and postgraduates engaged in Leather processing to better understand the developing trend of Leather industry in World, the history of leather industry, the knowledge of raw hides or skins, traditional technology of leather-making, clean production of leather manufacture, the reutilization of solid waste in the processing of leather manufacture and corresponding standards, etal, the book named "Eco Technology of Leather Manufacture" was written and published.

The book includes six chapters. Chapter 1, 2, 3, 4 are written by Professor Luo Jianxun, Jiaxing University. Chapter 5, 6 are written by vice-Professor Ma Hewei. The whole content of the book is guided by Professor Peng Biyu, Sichuan University, and Dr Juan C. Castell. In addition, some information were provided by professor He Youjie, sichuan university; Dr Jeffry Guthrie-strachan, JOAQUIN LINZOAIN ACEDO, Mr Zhang Jianwei, Mr AJ, Mr BAHNU, Mr Chen Fei, Mr Chen Zhengshuan, from TRUMPLER s. p. a, Mr CHUSOON, Mr Tang Peilong, Mr Xia Min from STAHL s. p. a and Dr Yi Yudan from Jiaxing University during the writing of the book. My thanks go to our friends together.

Because of the rapid development of technology of leather manufacture, incomplete information search, limitations of the author's personal level, et al, the book may has partly inappropriate aspects. Thus, it is hoped that the readers will kindly point out our errors.

Professor Luo Jianxun

Jiaxing University, December, 2022

Contents

Chapter 1 History of leather, introduction of leather industry ······ 1
1.1 History of leather ······ 1
1.1.1 Ancient times ······ 2
1.1.2 Middle Ages (Around the beginning of the 21st century BC to 221 BC) ······ 3
1.1.3 Lower ancient era (221 BC to 1840 AD) ······ 5
1.1.4 Modern society period ······ 7
1.2 Introduction of leather industry ······ 10
1.3 Distribution of leather industry in the world ······ 11
1.3.1 In Asia-Pacific ······ 11
1.3.2 In Europe ······ 13
1.3.3 In North America ······ 14
1.3.4 In Latin America ······ 15
1.3.5 In the Middle East and Africa ······ 16

Chapter 2 Basic knowledge of hides and skins ······ 18
2.1 Kinds of hides and skins ······ 18
2.1.1 Cattle hide ······ 18
2.1.2 Buffalo hide ······ 19
2.1.3 Yak hide ······ 20
2.1.4 Sheepskin ······ 20
2.1.5 Goatskin ······ 21
2.1.6 Pigskin ······ 21
2.1.7 Horse hide ······ 22
2.1.8 Reptile skin ······ 22
2.2 The structure of hides and skins ······ 23
2.2.1 The histological structure of hides and skins ······ 24
2.2.2 The chemical components of hides and skins ······ 26
2.2.3 Hair or wool and other components of hide or skin ······ 31
2.3 Curing and preservation of hides and skins ······ 33
2.3.1 Drying ······ 34
2.3.2 Salting ······ 35
2.3.3 Chilling and freezing ······ 36
2.3.4 Use of biocides ······ 36

2.3.5 Radiation curing ······ 37

Chapter 3 Traditional technology of leather manufacture ······ 38

3.1 Trimming and sorting ······ 38

3.2 Beam-house processing ······ 39

3.2.1 Soaking ······ 39

3.2.2 Degreasing ······ 43

3.2.3 Unhairing ······ 44

3.2.4 Liming ······ 48

3.2.5 Deliming ······ 50

3.2.6 Bating ······ 52

3.3 Tanning operation ······ 53

3.3.1 Pickling and depickling ······ 53

3.3.2 Tanning ······ 55

3.3.3 Chrome tanning ······ 56

3.3.4 Other mineral tanning ······ 61

3.3.5 Vegetable tanning ······ 63

3.3.6 Other tannages ······ 68

3.4 Dyehouse operation ······ 73

3.4.1 Rewetting process and neutralization ······ 73

3.4.2 Retanning process ······ 75

3.4.3 Dyeing process ······ 79

3.4.4 Fat-liquoring process ······ 82

3.5 Drying and softening operation ······ 85

3.5.1 The drying process ······ 85

3.5.2 The softening process ······ 87

3.6 Finishing ······ 90

3.6.1 Finishing materials ······ 91

3.6.2 Mechanism of forming film of film-forming materials ······ 94

3.6.3 Design of the finishing formulation and its properties ······ 96

3.6.4 Operations of finishing of the leather ······ 100

3.6.5 Other operations in the finishing process ······ 106

Chapter 4 Clean technology of leather manufacture ······ 110

4.1 Clean preservation technology of hide or skin ······ 110

4.1.1 Curing with less salt ······ 111

4.1.2 Curing with potassium chloride ······ 112

4.1.3 Curing with silicate ······ 113

4.2 Clean technology of beam-house operation ······ 113

4.2.1 Technologies of decreasing pollution in soaking process ………… 114
4.2.2 Technologies of decreasing pollution in degreasing process ………… 114
4.2.3 Technologies of decreasing pollution in unhairing and liming process ………… 115
4.2.4 Clean technologies of dispersing collagen fibers ………… 121
4.2.5 Technologies of pollution in deliming and process ………… 122
4.3 Clean technology of chrome tanning operation ………… 124
4.3.1 Clean technology of pickling process ………… 124
4.3.2 Clean technology of chrome tanning process ………… 129
4.4 Clean technology of chrome-free tanning process ………… 138
4.4.1 Tanning effect of single chrome-free tanning agent ………… 139
4.4.2 Combination tannage of two chrome-free tanning agents ………… 147
4.4.3 Complex combination tannage among more chrome-free tanning agents ………… 159
4.5 Clean technology of dyeing, retanning and fat-liquoring process ………… 161
4.5.1 Dyestuff and its high absorption technology ………… 161
4.5.2 Retanning agents and the high absorption technology ………… 163
4.5.3 Fat-liquoring agents and their high absorption technology ………… 163
4.6 Clean technology of finishing process ………… 165

Chapter 5 Resource utilization of tannery by-products ………… 169
5.1 Introduction ………… 169
5.2 Collagen process method ………… 169
5.3 Biogas ………… 170
5.3.1 Biogas from hair treatment ………… 170
5.3.2 Biogas process method ………… 170
5.4 Glue and gelatin ………… 171
5.4.1 Production of glue and gelatin ………… 171
5.4.2 Uses of gelatin ………… 171
5.4.3 Uses of glue ………… 172
5.4.4 Glue process method ………… 172
5.4.5 Gelatin from chrome shavings ………… 173
5.4.6 Case for treating fleshings by BLC ………… 173
5.5 Tallow recovery by mechanical means ………… 174
5.6 Chrome contained shavings ………… 175
5.6.1 Options for disposal ………… 175
5.6.2 Benefits ………… 175
5.7 Sludge ………… 176
5.7.1 Sludge processing ………… 176
5.7.2 Health and safety considerations ………… 177
5.7.3 Conclusions on methods processing sludge ………… 177

Chapter 6 Environmental protection and regulations ······ 179
6.1 Environmental protection related to leather processing ······ 179
6.1.1 Water pollution in leather production ······ 179
6.1.2 Environmental protection for leather production ······ 179
6.1.3 Regulations concerning substances contained in effluents ······ 180
6.1.4 Air pollution ······ 182
6.2 Critical substances in leather ······ 183

Reference ······ 186

Chapter 1 History of leather, introduction of leather industry

1.1 History of leather

1-1

Leather is one of the earliest products and Leather industry is one of old and modern industries. Leather industry acts as a traditional industry and keeps up with human civilization whose history is the extremely important component of the history of human beings. Leather, double face and their products have become one of daily necessities since human being appeared.

Many thousands of years ago, primitive man began to make leather (Figure 1.1), which would have been one of the very first manufacturing industries. Animals were hunted and killed for food, but the skin had to be removed from the animals before they were eaten. Sharp flints were probably used to peel the skin away from the carcass. The skins would then have been worn for warmth and protection from the elements and probably wrapped around the feet. But the skins would soon begin to decompose and rot away. So our ancestors realized that drying the skins would preserve them. However, the dried skins were very hard, inflexible and uncomfortable material. When the skins would have been done by rubbing with fats, brains or bone marrow, the skins would have to be softened and wearable. In addition, the fat, brains or bone marrow can help prevent the skins getting wet, thereby facilitating long-term storage and enhancing blending and flexing properties of skins. Therefore, the skins can be softer and more pliable for wearers.

Then primitive man found that the skins can prevent insect and corrosion after they had been smoked from lighting woods. As shown in Figure 1.2 and Figure 1.3. Lately, a liquid containing vegetable extracts was made by using water, various barks, leaves and berries. When the skins were immersed in it, the skins became rot-resistant and considerably softer than the dried skin. The active agents in this liquid are called tannin. This was probably one of the first methods of tanning leather. This process of tanning skin spread and was improved. By Roman times, armour, water containers, belts, straps, tents, boats, etc were regularly being made from leather. By

Figure 1.1 Primitive man and obtained prey

the middle times, things began to be very well organized. Tanneries were set up and mainly concentrated into special areas which have good sources of materials such as a supply of hides and skins, plenty of water, lime for softening and assisting with hair removal and plenty of trees for the extraction of tannin from the bark.

Figure 1.2 Primitive men were making leather

Figure 1.3 Clothing made of animal skins of primitive man

Take the development of leather industry in China as an instance, the history of leather industry is introduced which experiences ancient times (3,000,000 BC to 10,000 BC), ancient times (Early 26th century BC to early 21st century BC), Middle Ages (Around the beginning of the 21st century BC to 221 BC), lower Ancient Period (221 BC to 1840 AD) and modern society.

1.1.1 Ancient times

1.1.1.1 Remote antiquity (3,000,000 BC to 10,000 BC)

Animal grease, brain and bone marrow are used to smear on the flesh side of the leather and make it soft and made it soft and wearable by mechanical actions such as rubbing, etc. Subsequently, it was discovered that fuming the hide or skin with the smoke from burning wood could prevent insects and corrosion. Later, the hide or skin was treated by the immersion solution of the bark. Dried hide or skins can be neither shrunk nor rotted which can be made into a variety of soft and tough, long-term preservation appliances.

1.1.1.2 The Remote Ages (Early 26th century BC to early 21st century BC)

In this period (Primitive society), the technology of leather - making had rapid development. Primitive leather industry began to rise. Between 10,000 BC and 2,500 BC, human beings invented the method of "unhairing methods by sweating". After the invention was used in leather industry, human beings entered into the times between leather and double face. It was said that there was properly leather shoes, garnment, trousers in Huangdi's era (about five to six thousands of years ago), which is recorded in "*The History of Industry of Modern China*" "*Records of the Grand Historian: Age of Emperors*".

1.1.2 Middle Ages (Around the beginning of the 21st century BC to 221 BC)

1.1.2.1 Xia and Shang Dynasties (2,200 BC to 1,100 BC)

Human beings invented the unhairing methods of lime-liquid which was later improved to the unhairing process of lime-alkali by adding sodium sulfide into the lime solution. At the same time, there were all kinds of shoes (leather, cotton, cloth, grass). The cattle hide ankle boots (Figure 1.4) found in the ancient tombs in the Tiepan River area north of Lop nur in Xinjiang are the oldest boots left in the world and are known as the "World's first boots".

Figure 1.4 Women's sheepskin boots unearthed from ancient Loulan tombs

The word "qiu" appears in the oracle bone inscriptions and Chinese bronze inscriptions, as shown in Figure 1.5, and the "leather" word appears on the Rong Revolutionary Cauldron. At that time, leather was used as leather armor which was used to protect themselves in the war acting as unique defensive weapons.

甲骨文	金文		篆书		隶书	
裘	裘	革	硝	矾	硝	矾

Figure 1.5 "Qiu" "Ge" "Xiao" "Fan" writed in oracle bone script, Chinese bronze inscriptions, seal charactes and clerical script

1.1.2.2 Shang and Zhou Periods (1,600 BC to 256 BC)

The Zhou Dynasty established five official positions of "gold, jade, leather, art and stone" and officials were in charge of daily necessities of the people, as shown in Figure 1.6. "Leather" was an official position who is specially in charge of animal skin products which is recorded in *The History of Industry of Modern China*. Leather shoes were already popular in Zhou Dynasty. In the Western Zhou Dynasty, there were tanners who used leather to make military shoes, which was collectively one of *hundreds of works*. At that time, people had mastered the technology of leather-making which could make animal skin into soft double face leather clothes. There is a sentence "The sheep double face is carefree, and the fox double face in court" in *The Book of Songs*.

1.1.2.3 The Spring and Autumn Period (770 BC to 221 BC)

Important handicraft industries such as the technologies of leather manufacture, Leather armor,

Figure 1.6 The five official positions of "gold, jade, skin, workmanship and stone" recorded on oracle bone inscriptions

Leather drum, etc were recorded in *Kao Gong Ji*, which is one of the earliest existing ancient literatures. It is thus clear that leather industry was relatively developed. The requirements of the technology of leather-making and its inspection methods were described in the "Pao Ren's Affairs" which was recorded of *Kao Gong Ji*. The animal hides or skins were processed into leather with vegetable tannage and aluminum tannage (Figure 1.7). Thus the quality of leather products is greatly improved. In the "*Wearing Hu dress and shooting on horse*" put forward by Zhao Wuling Emperor (Figure 1.8), leather boots were used in the army.

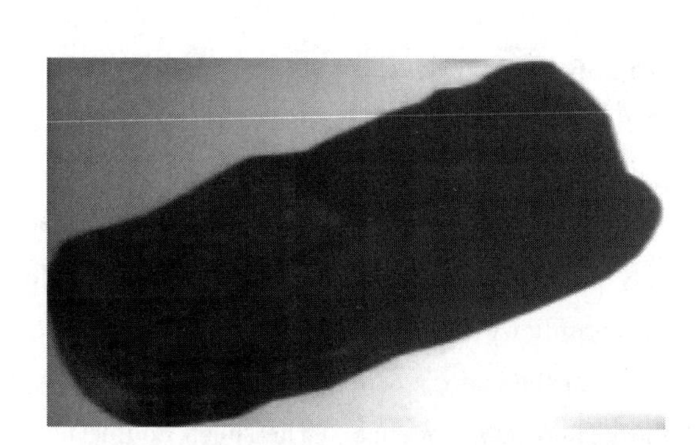

Figure 1.7 Cattle-hide shoes during the Warring States Period

Figure 1.8 Zhao Wuling Emperor's "*Wearing Hu dress and shooting on horse*"

1.1.3 Lower ancient era (221 BC to 1840 AD)

During this period (the feudal society), leather and its articles were widely used and the leather industry was rapidly developed. Brilliant achievements in leather industry were created in the Han Dynasty. Leather industry was popularized and improved in Wei, Jin, Southern and Northern Dynasties, and reached its peak in the Sui and Tang Dynasties, followed by the Song Dynasty. Leather industry formed the national style in the Liao, Jin and Xi Xia periods, reached a new height in the Yuan Dynasty, took on a new look in the Ming Dynasty, and reached an extremely prosperous situation in the Qing Dynasty. However, due to the self-complacency of small-scale peasant economy, technologies of leather-making remained stagnant and the traditional methods were still adopted in the production such as mirabilite tanning, smoking tanning, aluminum tanning and tanning with the immersion solution of the bark. The scale of the production also stayed at the stage of a manual workshop. These led to the technology of China's leather production far lagging behind that of advanced western countries.

1.1.3.1 Qin and Han Dynasties (221 BC to 220 AD)

During the reign of Emperor Qin Shihuang, skilled leather dyeing techniques were developed and leather armor was widely used in the military. In the Western Han Dynasty, leather and fur industries were developed and became a symbol of the feudal hierarchy. At that time, the Hefei area, Anhui province was the distribution center of hide or skins. During the reign of Emperor Wudi of the Han Dynasty, white deer skin was used as currency. At the same time, donkey skin and cow skin were used to make shadow puppets, and sheepskin rafts are used to cross-rivers in Hetao regions of Gansu, Shaanxi, Ningxia and Inner Mongolia. During Western Han Dynasty, China's double face leathers were exported and exchanged with the royal family as gifts, which is recorded in *A History of Science and Technology* in *China by Joseph Needham* and *The Oriental History of the Han Dynasty*. In Sichuan, leather was used to make saddles, leather belts, ornaments of chariots, winter clothes, etc. During salt-making and smelting, cowhide was used for picking brine, and blowing (Figure 1.9).

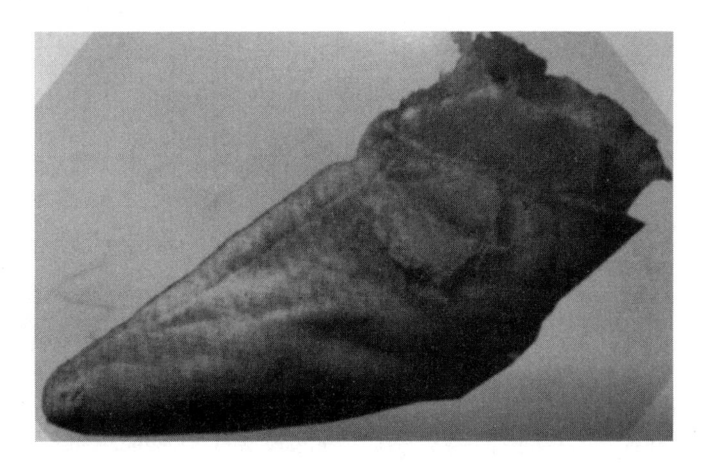

Figure 1.9 Shoes made of roe and muntjac skins during the Qin and Han Dynasties

1.1.3.2 Wei Jin North-south Dynasties (265 AD-581 AD)

Leather boots and shoes made by animal skin were commonly used by northern nationalities. People living in Guangdong knew how to use cowhide to make clothes, boots, tents, carpets and other articles of daily essentials. Leather products in Lingnan, which was known as "*southern goods*", were sold as gifts or commodities to the north.

1.1.3.3 Sui and Tang Dynasties (581 AD-907 AD)

During the Sui Dynasty, leather boots were popular, especially the famous "*Liu-he boots*" made of six pieces of leather (east, west, south, north, heaven and earth). At the same time, wearing leather saguaro or leather caps and black leather shoes had formed a system. During the Tang Dynasty, fur and leather production had reached a certain scale. Liu-he boots, boots, ankle boots and pointed ankle boots made by colored leather were very popular. In addition, the processing of natural double face leather such as mink, fox, roe deer, bear, tiger, etc was the main industry in Jilin, northeast district. Mink had become a tribute, commercial commodity (Figure 1.10).

Figure 1.10 An ancient cap made of deer hide leather

1.1.3.4 Song Dynasty (960 AD-1279 AD)

The development of gunpowder, compass and printing technology also promoted the rapid development of leather industry. In the early Song Dynasty, in order to meet the needs of the military, the state vigorously developed the leather industry by establishing a large number of tanneries and being administrated by the military supervision. In addition, livestock, handicrafts of fur and leather were introduced from the north to the south as the nomadic minorities in the north migrated to the south, which promoted the development of the breeding industry and leather industry in the south.

1.1.3.5 Yuan Dynasty (1206 AD-1368 AD)

The period was the heyday of leather industry in ancient times. Relatively mature vegetable tannage was widely used in the production of leather. The hard buffalo hide or cowhide leather was used to make defensive armor. In addition, long boots, short boots, square boots and the inflexible "Mongolian boots" were used for military boots. At the same time, fur coats made of sable, white fox, black fox, lynx and other double face leathers gradually turned from the royal nobility to the public (Figure 1.11).

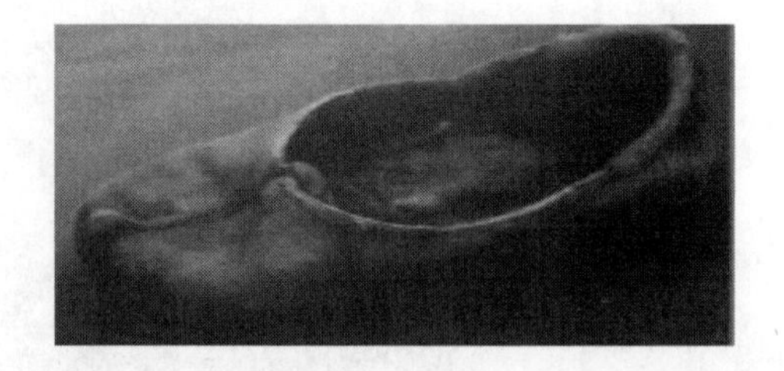

Figure 1.11 Leather shoes in the Yuan Dynasty

1.1.3.6 Ming Dynasty (1368–1644)

The sentence "The hair of chamois is removed, tanned by mirabilite and further made into garments. They can resist wind and protect the body. Socks and boots made by leather are better" is recorded in *Tiangong Kaiwu*, which has depicted the methods of leather processing and the excellent properties of leather fur products . The technology of leather – making in the Ming Dynasty mainly consisted of mirabilite tanning and smoking tanning. At that time, the imperial court set up *the Bureau of Fur and Horses* which specifically administrated the leather industry including the animal hides or skins and tanneries. In addition, the Ming Dynasty advocated the footwear culture of the Han, Tang and Song Dynasties which brought the footwear culture to maturity. Ankle boots, boots made of moire head leather boots were used in the army. Double face leather clothing had become the favorite clothing of aristocrats and literati class. Leather industry in Ji–ning, Lin–qing, Ding–tao, Liao–cheng and other places, Shandong province is flourishing. Especially green hedgehog skin produced in Ji – ning and sheep shear cashmere produced in Ding – tao, etc had long enjoyed a good reputation. Leather industries in Xin–ji, Hebei province, southwest of Zhejiang, Hu–nan (making smoked leather, white leather, cow leather, saddle, scabbard, oil shoes, clogs, etc.), minority areas in Sichuan (oil tanned leather), Lanzhou (making clothes, hats, shoes, bags, saddles, scabbard, water bags, cattle and sheepskin rafts, etc) and other places were more prosperous than be fore.

1.1.3.7 Qing Dynasty (1616–1840)

It is the prosperous period of China's leather industry. Double face leather clothes became the fashion of the aristocracy which greatly stimulated the development of leather industry. Quartermaster equipment made of patent leather was also very popular. At that time, leather industry in Da–tong and Jiao–cheng areas, Shanxi entered a prosperous period.

1.1.4 Modern society period

1.1.4.1 Development of Leather industry in modern China (1841–1949)

Since the 1830s, leather industry had spread all over China and formed a number of concentrated production areas which were mainly distributed in Shanghai, Beijing, Tianjin, Hebei, Liaoning, Jilin, Shanxi, Shandong, Hubei and other places. Shanghai is the birthplace of the modern leather industry in China. The imperial city of Beijing had the reputation of "*Double face leather is the best around the world*". Tianjin had the earliest tannery in modern China. Zhangjiakou had a reputation of "north leather capital" and their fine leather, knife Were Fengtian were famous in the world.

In 1898, China's first modern tannery–Tianjin Beiyang Tannery was established in Tianjin which adopted the modern tanning technology (In 1858, the application of chrome tanning technology had brought leather manufacture into the "modern" era.) and machinery (Termometer, hydrometer and drums were used during the leather – making. The splitting machine was invented in 1809 by Boyden. Mechanical de–hairing and fleshing machine were put to use in Leather processing in 1840). The technology of leather–making and related chemical materials were the most advanced in Europe at the end of the 19th century.

In 1876, Shen Binggen founded China's first leather shoe factory in Shanghai. In 1919, the Beijing Leather Shoe Factory established on Guangdong Road in Shanghai firstly used machines to produce leather shoes. At the beginning of the War of Resistance Against Japan, there were more than 200 leather shoe factories in Shanghai. However, there were more than 800 leather shoe factories after the victory of the War.

However, due to the weak national strength and the semi-colonial and semi-feudal society, China's leather industry was lag in technology which was controlled by foreign companies. The equipment and chemical materials of tanneries depended on foreign countries.

1.1.4.2 Development of modern leather industry in China (1949 AD-present)

- Restoration period (1949-1957):

The leather industry gradually took the socialist road by some policies and relevant measures formulated by the state council and the Ministry of Light industry. In order to expand the source of raw materials and effectively ensure supply, pigskin was used for the preparation of leather. Initial results of restoration and transformation have been achieved, and the production level has been greatly improved. In 1957, the output of light leather was 14.34 million square meters, an increase of 290% than that in 1952. The production of shoes was 30 million pairs, an increase of 250% compared with that in 1952, and the export trend was formed. During this period, China's leather industry fully resumed production, and the quantity, variety and quality of leather and leather products were improved to meet the basic supply and social requirements of the military, industry and people's livelihood.

- Initial construction period (1958-1977):

China's leather industry appeared in a depressed state, but always maintained growth and the production capacity greatly improved and the product varieties were gradually completed because it experienced the "Great Leap Forward", three years of natural disasters, three years of adjustment, the "Cultural Revolution" and the test of historical change. With the progress of science and technology, China's leather products were increasingly rich. Light leather products such as cattle leather, sheep leather, ball leather, split leather, etc have been preliminarily serialized and heavy leather such as bottom leather, industrial leather, leather fittings, belt leather, etc can meet the requirements of production and life. The output of leather products increased significantly. In 1977, the output was 25.19 million pieces of leather (coverted to cattlehide) and more than 91.19 million pairs of leather shoes (6.39 million pairs were exported). At the same time, pigskin leather obtained greater development, and has become the main raw material leather in China. Moreover, new products based on pigskin leather were constantly emerging.

- The early stage of reform and opening up (1978-1987):

Leather industry developed rapidly on the basis of reconstruction and initial construction. The level of mechanization in leather industry was steadily improved. New changes emerge in economic structure, that is, to form a self-improving industrial system with leather, leather goods, fur and fur products as the main industries and leather chemicals, leather machinery, leather hardware, shoe materials as supporting industries which had laid a solid foundation for the development of the industry in the later period. The production and technology of pigskin leather achieved great development. In 1987, the output of pigskin leather reached 8253 pieces and there were many styles. The degree of mechanization of the industry had reached 50%–70% and it was gradually transitioning to mechanized and semi–mechanized production mode. Leather chemical industry and leather machinery Gradually shifted from relying mainly on imports to relying mainly on domestic self–sufficiency. Industry systems change so that township private enterprise appeared.

• Rapid development period (1988–1997):

With the deepening of the reform and opening up, the establishment of the market economy system and the shift of the center of world leather, China's leather industry has achieved rapid development and great changes emerged in improving product quality, expanding export to earn foreign exchange and adapting to the market economy, etc. The level of production and the ability to earn foreign exchange in the whole industry have been greatly improved, the quality of the main products, total imports and exports are among the best in the world so that China had become the processing center and sales center of world leather and initially established the status of China's world leather power. At this time, leather industry owned a complete industrial chain, technology, product quality which was geared to international standards. The economic structure of the industry had undergone profound changes. State – owned enterprises account for only 7% of the total, while the rest were collective private enterprises and foreign–funded enterprises, and a number of excellent enterprises have emerged. Cluster production and specialized markets became the new platform for the development of leather industry. The certification trademark "Genuine leather mark" registered by the China Leather Association became the brand platform of the industry in the market economy.

• Broad prosperity period (1998–2010):

During the process of making great efforts to realize strategic objectives of the "second venture", leather industry withstood the test of the EU anti–dumping and the international financial crisis, also seized the opportunity to enter the international market by joining the WTO, the level of production, export in leather manufacture constantly rose. Remarkable achievements were obtained in the adjustment of industrial structure and brand construction. Leather industry in China presented comprehensive prosperities.

During this period, industrial clusters and characteristic areas from raw materials, processing to sales and services had been formed which steppted up the development of the local economy. Characteristic areas were initially formed such as shoe–making factories in Wen–zhou, Zhejiang Province, Cheng–du, Sichuan Province, Bi–shan, Chongqing, Guangdong, and Fujian, processing of leathers and garment leathers in Hai–ning, Zhejiang, Xin–ji, Hebei, leather articles in Shi–ling town, Guangzhou, Quan–zhou, Fujian, Bai–gou Hebei, processing of double face leather in Tongxiang city, Zhejiang Province, Su–ning, Hebei, Sang–Po, Henan, etc. Because of labor prices and other factors, the distribution of characteristic areas was gradually reasonable and the industrial transfer is accelerating. More than 100 domestic and international brands emerged and entered the international market such as "China leather Shoes King", "China well–known leather shoes", "China genius leather king", etc. The outdated production capacity of less than 20 million standard sheets and technologies including azo dyes, formaldehyde, and solvent based finishing materials are gradually eliminated. Innovation and industrialization of clean technology of leather manufacture had achieved remarkable effects.

The Successful financing and listing of Hai-ning China Leather City, Yue-hai Leather and other successful financing and listing enterprises, etc enhance the social influence and popularity of the companies. In 2009, China's output of cattlehide leather, sheepskin leather accounted for 14% and 27% of the world. The output of pigskin leather ranked first in the world. Thus the development of leather industy firmly established the status of world's leather power.

- Transformation and upgrading period (2010 to present):

According to the requirement of the "innovation, coordination, green, open, sharing" new development concept, great efforts have been made in clean technologies of leather production, environment-friendly chemicals, the treatment and recycling of the effluent and solid waste, etc in China's leather industry and the road-map of water-saving and emissions-reducing of technology in leather industry is put forward which is being studied.

1.2 Introduction of leather industry

As mentioned earlier, the history of leather production goes right back to prehistoric times, when primitive methods were developed for treating animal hides and skins so that they could be used for clothing to protect people from the elemments. Therefore, the aim of leather industry is to convert raw hides and skins into leather which can be made for all kinds of leather goods. Just like the raw hides and skins, the leather will not decay and last for hundreds of years. Therefore the technology of leather manufacture is very important for the quality of leather and relevant leather goods, which should meet customers' requirements.

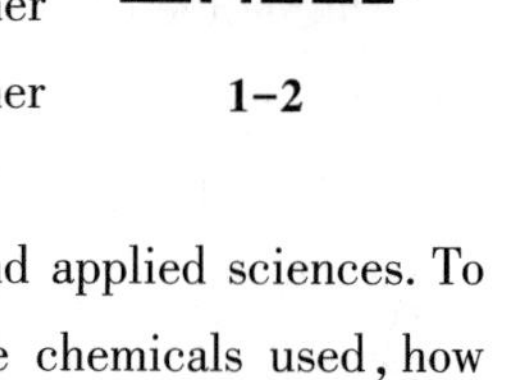

1-2

The leather technologist has become familiar with a wide range of pure and applied sciences. To produce high-quality leather, technologists must understand the nature of the chemicals used, how they react, the means of controlling this reactivity, and the methods of testing and analyzing the finished product. With this knowledge as a basis, tanners must become familiar with all the practical processes and machinery operations that are necessary to produce leather.

Leather manufacture involves beam-house operations, tanning operations, post-tanning and finishing operations.

The aim of beam-house operations is to remove the non-collagenous substances such as fur, fats, non-collagen protein, etc, and loose the collagen including soaking, fleshing, degreasing, unhairing, liming splitting, deliming and bating processes. The aim of tanning operations is to convert the treated hides and skins into leather which overcomes revolutionary changes including pickling, depickling and tanning processes. In the tanning process, the collagen fiber of the hide and skin is stabilized by the tanning agents so that the hide and skin is no longer susceptible to putrefaction, and the tanned hide and skin is called wet-blue or wet-white. They can be traded as intermediate products. However, if wet-blues are to be used to manufacture consumer products, they further need to be processed and finished.

Then post-tanning and finishing operations are to further process the shaved wet-blue or white-blue in order to obtain fashionable, functional leather including rewetting, retanning, neutralizing, dying, fat-liquoring, drying, softening and finishing processes. According to the desired leather the wet-blues or wet-whites are retanned to improve the handle, dyed with water-soluble dye-stuffs to produce even colour over the leather surface, fat-liquored to lubricate the fiber and finally dried. After drying, the leather may be referred to as crust, which is a tradable intermediate product.

Finishing operations are to give the leather as thin a finish as possible without harming the thermatural characteristics of leather, such as its look and its ability to breathe. By grounding, coating, seasoning, embossing and ironing, the leather will have a shiny or matt, single or multi-colored, smooth or clearly grained surface. The overall objective of finishing is to improve the appearance of the leather and to provide the appropriate characteristics in terms of color, gloss and handle.

Operations carried out in the beam house, the tanning area and the post-tanning areas are often referred to as wet processing, as they are performed in processing vessels (paddle or drum) filled with water to which die necessary chemicals are added to produce the desired reaction. After post-tanning the leather is dried and subsequent operations are referred to as dry processing.

1.3 Distribution of leather industry in the world

Leather industry is a traditional, characteristic industry and is widely distributed in the world, such as Asia, Africa, Europe, America, etc. Advances in production and technology have resulted in a complex global commodity market.

Leather is a stretchable and durable material made by treating hides and skins of various animals, such as buffaloes, sheep, cattle, goats, hogs, horses, and camels. It is used to manufacture gloves, bags, clothes, watches, footwear, saddles, harnesses, and furniture due to its enhanced strength, durability, and flexibility. It is long-lasting, repairable, comfortable, and resistant to dust, fire, scratch, and water. As a result, it is increasingly being utilized in the manufacturing of products like automotive upholstery, sports equipment, and apparel.

Advances in production and technology have resulted in a complex global commodity market. The global leather goods market reached a value of US $350 billion in 2021. Looking forward, IMARC Group (a famous consulting company) expects the market to reach US $490.2 billion by 2027, exhibiting an annual growth rate of 6.1% during 2022-2027. Developing countries have become the world's largest producers of leather, especially in Asia, mainly due to cost competitiveness and less strict environmental regulations. The location of leather industry in the world was exhibited in Figure 12.

1.3.1 In Asia-Pacific

Numerous leather manufacturers in Asia are engaged in the production of leather and leather articles. Leather coats, jackets, skirts, trousers, belts, gloves, bags, shoes, and wallets made in Asia are famous all over the world. Nine out of ten pairs of shoes in the world are produced in Asia, accounting in 87% of the total world footwear production. Despite the impacts of the COVID-19 pandemic, the geographical distribution of leather production in the world has been intact since 2022.

China is the largest country in leather industry and accounted for 23% of world leather production in 2022. Today, there are over 120 scaled tanneries, 3,800 shoe-making factories and over 8,000 leather clothing enterprises in China. In 2022, the leather processing capacity was about 5,290 million sq. ft. The leather clothing production was over 21.3 million pieces, and the leather shoe output had been over 1.7 billion pairs. The leather manufacturers are mainly located in Zhejiang, Hebei, Guangdong, Fujian and Henan. Haining is considered as the most important production and trade center of the leather garment industry.

The leather industry has a significant role in China's foreign trade. Its export value was about US $72 billion in 2021. The most important export item is leather shoes and leather wear products. Its annual import value is about US$50 billion. The most important import item is raw hides. United States is the most important market of China's shoe, accounting for about 30% of China's shoe exports. Europe was the second largest export market. In recent years, China has already been the country in the world with the largest output and export of leather products.

In India, the leather industry holds a prominent place in the economy. This sector is known for its consistency in high export earnings and it is among the top ten foreign exchange earners for the Country. The export of footwear, leather and leather products from India reached a value of US $5.07 billion during 2019-2020.

The leather industry is bestowed with an affluence of raw materials as India is endowed with 20% of the world cattle & buffalo and 11% of the world goat & sheep population. Added to this were the strengths of skilled manpower, increasing industry compliance with international environmental standards, and the dedicated support of the allied industries. In 2015, the leather processing capacity was about 1,516 million sq. ft, accounting for 6.3% of world leather production. The leather industry is an employment intensive sector, providing jobs to about 4.42 million people, mostly from the weaker sections of the society.

India is the second largest producer of footwear and leather garments in the world, and the second largest exporter of leather garments and the third largest exporter of Saddlery & Harness. The major production centers for footwear, leather and leather products are mainly located in three States of Tamil Nadu, West Bengal and Uttar Pradesh. Of the 1,500 tanneries in India, Tamil Nadu accounts for 50%, West Bengal 20%, and Uttar Pradesh 10%. The other important states are Maharastra, Andhra Pradesh and Punjab. Looking from the angle of the scale of operations, the Indian leather industry largely exists in the small sector with a share of 90% in the total number of tanneries. It is apparent that the small scale sector accounts for large processing capacity ranging from 70% to 80% for different leathers.

Vietnam's leather shoe industry has developed very quickly and is considered one of the driving forces of the Vietnamese economy. With about 240 operating businesses, Vietnam's leather and footwear industry is a key export industry, attracting about 500,000 employees. The US and European market accounts for 70%

of Vietnam's leather exports. According to the Vietnam Leather, Footwear and Handbag Association, the exports of leather and footwear products were estimated to be US $16.5 billion in 2020. Most of the input materials used for leather processing were mainly supplied from China.

In Bangladesh, leather is one of the oldest industries. Exporting 10 percent of the global demand for leather, Bangladesh's leather industry has become the country's second-largest source of foreign exchange. Having a favorable environment for raising and nurturing animals, Bangladesh has 2% of the total livestock population in the world. In 2019, Bangladesh produced 350 million square feet of leather, of which 80 percent was exported. By 2020, 378 million pairs of shoes were manufactured in Bangladesh every year, which promoted Bangladesh to be the 8th largest footwear producer in the world with a 2.1% share of Global Shoe Productions. Currently, Bangladesh's leather industry is gaining the capacity to produce processed raw leather and leather products in sustainable ways.

South Korea is a net importer of leather. South Korea mainly imports raw hides and skins (Fresh or Preserved) from the US, once accounting for 40% of the total US raw hides. In 2015, the leather processing capacity was about 1,144 million sq. ft, accounting for 4.8% of world leather production. However, the annual growth of South Korea leather in value was-24% per year between 2015 to 2019, with the annual growth in quantity during the same period being-62%, per annum. Especially in 2020, the leather inndustry was severely Hit Hard by Covid-19, leading to heavy losses and the largest drop in sales.

1.3.2 In Europe

Europe is an important player in the international leather trade. With about 25% of the world's leather production and one of the largest and most dynamic consumer markets for leather articles, Europe stands out as the leading force in international business circles in relation to leather and tanning.

European tanneries are always showing their competitiveness in the global market. Their products are renowned and appreciated by manufacturers worldwide for the quality and fashionable designs. European leathers are exported all over the world to satisfy the highest standards, the most stringent ecological regulations and the increasing expectations of quality aware consumers.

Tanners in Europe have a long tradition of producing all kinds of leather, from bovine and calf leather to sheep and goat leather, from sole and exotic specialties to double-face garment leather. Their expertise contributes to the success of leading footwear, garment, furniture and leather goods manufacturers. This solid experience and the outstanding know-how of European tanners and dressers is displayed at major international fairs. All this explains the continuously strong demand for their products on international markets.

Snail and medium-sized companies predominate in the European tanning sector. The consequent flexibility, adaptability and quick response to demand constitute one of the industry's most important assets. Larger companies, however, excel thanks to their capacity to be at the forefront of technological developments and to constitute reliable partners in global business activities. Technological leadership, fashion, design, quality, excellent materials and the sense of service to the customer are all factors contributing to the strength of European tanners. Continuing modernization combined with investment in training, environmental infrastructure, R&D and export promotion allows the industry's operators to look to the future with confidence.

In the last decade, European finished leather exports to other regions in the world have experienced extraordinary development. Manufacturers in new and emerging markets are particularly keen to develop trade relations with European tanners and to work closely with them. Despite the leather goods market has witnessed a sharp decline in demand due to the COVID-19 outbreak, as the export of leather goods products in Italy decreased by 13%, from 12.4 million in 2019 to 9.6 million in 2020, the market has been slowly recovering since the lockdown opened in Europe.

1.3.3 In North America

The United States produces the largest cattle hides annually. According to LHCA (The Leather and Hide Council of America), the US exported more than $1.4 billion in combined cattle hides, pigskins and semi-processed leather products in 2021. China was the largest buyer of salted cattle hides, with imports valued at more than $584 million. China also edged out Italy as the single largest destination for wet blue cattle hides, with imports valued at more than $131 million in 2021. Other large export markets included South Korea, Mexico, Thailand, and Vietnam. What has to be indicated is that most of these exports come back to the US in the form of finished shoes and other products. Leather footwear imports accounted for almost 0.5 billion pairs in 2021, out of which 30% came from China and 34% from Vietnam.

In the past decades, the American leather industry has registered significant losses. Almost every sector in the leather industry has suffered. The leather footwear segment is particularly touched by the invasion of cheap pairs of shoes. In 1997, there was a peak in the number of US hides available for making leather, and since then this category has remained decline (as shown in the Figure 1.12). The leather tanning industry continues to experience low prices with the Asian invasion. As the US leather sector feature is mainly that of being wet-salted and wet-blue supplier, the strategy followed by the tanners is that of targeting niche markets and high value costumers rather than focusing on mass consumers.

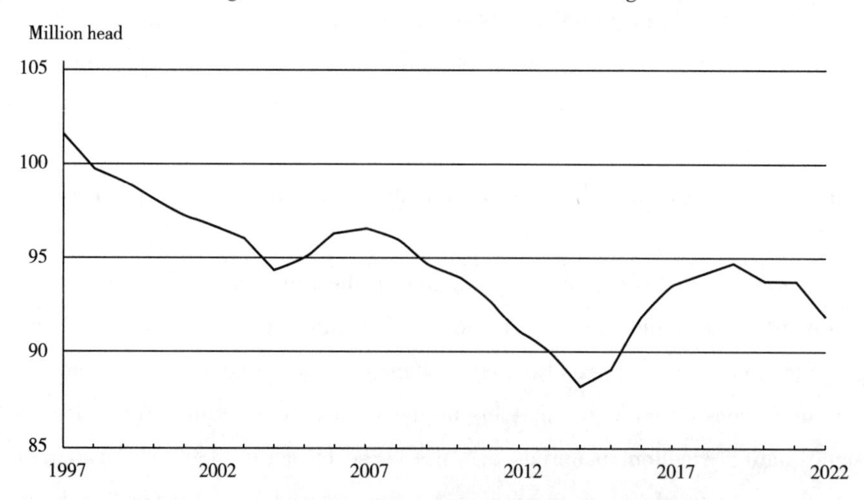

Figure 1.12 All cattle and calves inventory of the US

To gain competitiveness, many US businessmen have transferred their operations from US territory to Mexico or China in order to lower production costs. Besides, other than labor costs, one of the greatest inconveniences of tanning in the US is the strict environmental regulation. Making tanneries more environmentally friendly is still a matter to take into account. Despite these challenges, the US industry is poised to again contend with changing global demand and price fluctuations that seek to mischaracterize leather's sustainability credentials.

The Mexican leather industry is geographically very compartmentalized and mostly located in the state of Guanajuato. About 80% of leather production is in Leon - Mexico's fourth largest city. The average tannery does not have very high technology, and the average age of the machinery varies from 15-18 years. Since Mexico does not produce tanning or footwear machinery, all must be imported at a high cost. Also, most of the machinery imported is used equipment.

In 2014, there were approximately 650 tanneries throughout the country, of which 80% could be considered micro tanneries, and totally produced almost fifty thousand of hides each day. They mostly concentrate their production into four categories: footwear, designer products, furniture, and automobile upholstery. Only 20 companies produce and export auto and upholstery leather. Most tanneries produce shoe upper leather. A growing number of tanneries are processing designer products and furniture for both domestic use and export. These products, especially furniture, require better and more refined leather.

Mexico's domestic market is large but poor, and since per capita income is low, experts do not foresee significant growth in the consumption of footwear in the near future. Consequently, leather manufacturers are looking abroad, mostly to the United States, for new markets.

1.3.4 In Latin America

Brazil is one of the major global leather producers. In 2015, the leather processing capacity was about 2260 million sq. ft, accounting for 9.4% of world leather production. Brazilian leather is much appreciated abroad, and is present in more than 35 different countries, including Italy, China and the United States, with shipments of 34 million units a year. This position is explained by various comparative advantages, beginning with the abundant supply of raw materials. After all, the country has the world's largest commercial beef cattle herd, estimated at over 190 million heads of cattle.

The leather production chain has been one of the major driving forces of the Brazilian economy. The industrial complex is comprised of the tanneries, shoes, components, machinery, and leather artifacts sectors. The activity generates a turnover of more than US $21 billion per year, is comprised of 10,000 industries, and employs over 500,000 people.

For Argentina, because of more than 55 million head of bovine cattle and a sturdy, traditional, professional, export-oriented and well-known tanning industry, it has always been a factor in the global picture of the leather industry. In 2015, the leather processing capacity was about 804 million sq. ft, accounting for 3.4% of world leather production. The leather industry employs 60,000 people and even though the decrease of the transformed goods, the sector appears to be strong and willing to survive the international crisis due to COVID−19. The three largest buyers of Argentina's leather goods are China, America and Italy.

Argentine leather in wet−blue, crust and finished state has been exported for a total value of approximately \$2.0 billion during 2015, of which a fair estimate is that 50% is car and furniture leather, and the rest is for shoes and leather goods.

However, for the past 40 years, Argentina has had a fluctuating export tax on bovine wet−salted and wet−blue hides due to the state of the export tax regulation. It leads to a 20% decrease of cattle population and the contraction of 30% of the industry with the disappearance of some great tanneries.

1.3.5 In the Middle East and Africa

In recent years, the leather industry in the Middle East has developed rapidly. Among them, the leather trade of the United Arab Emirates in the Gulf area is relatively large, about US\$2 billion a year, and the total import and export trade of leather goods in Saudi Arabia in 2013 was US\$1 billion 100 million. Dubai has gradually become the main pit center for the Sultan, Kenya and Ethiopia. In the United Arab Emirates, a tannery for camel leather has 20 years of history and is one of the most advanced leather making equipment manufacturers in the world. Camel skin is more expensive, but it provides another kind of business outlet for local camel farms.

For Turkey, leather has a legacy of hundreds of years and has been flourishing. Today, Turkey is one of the biggest producers of high quality leather products in the world. It is mainly known for processing sheep and goat leather. Besides, 22% of the world's small cattle leather processing and production is obtained from Turkey. The leather processing capacity was about 529 million sq. ft in 2015, accounting for 2.2% of world leather production.

Turkey has become the focal point for leather materials, manufacturing and technology in the world. Turkey has 13 industrial leather zones which use modern technology and produce high quality products. Turkey holds the fourth place after China, Italy and India for being the biggest producer of leather. Besides, Turkey is the world leader in the production of fur with a processing capacity of 80 million units yearly. Tuzla has become the biggest industrial leather sector.

Turkish leather industry has laid down its action plan for the next decade. It has earned laurels for its leather exports and now it is keen to showcase Turkey's expertise in leather design on a global platform. Turkey is rocking the global leather fashion industry with its designing skills.

African countries have an abundant and renewable resource base in Africa's large population of cows, sheep and goats. It has, however, major obstacles to overcome to realize this potential. The main problem lies in the collection and processing of its rich supply of hides and skins, because the predominant practice is to keep animals for their meat and not for their hides and skins.

This continent has 20% of the world's cattle, sheep and goats, but produces only 14.9% of the world output of hides and skins. They have 10% of the world's cattle but produce only 4.5% of bovine hides. In past decades their exports of hides and skins fell from 4% to 2%, and their tanning capacity from 9% to 7%. At a time when other developing countries have substantially increased their share of world footwear production in relation to developed countries, African countries have shown only a modest increase. Import penetration of their domestic leather footwear markets by other developing countries is estimated at 73%.

This gap between resources and production shows the considerable potential of the African leather industry. Reducing this gap is especially critical in an important strategic sector for the economic and industrial development of many African countries. Not only does this sector have an excellent and renewable resource base, but it is also labor intensive with the potential to be a major source of employment.

Leather industry has been an important strategic sector for the economic and industrial development of many African countries. In the eight countries, leather and shoe manufacturing already provides 4% to 5% of total industrial employment, with contributions to manufacturing value added of 2.9% in Egypt, 8.3% in Tunisia and 74% in Ethiopia, where the cattle population is the highest in Africa, and close to 1% in the remaining five countries. Clearly, the realization of the African leather industry's potential would bring significant economic gains to the continent.

Questions

(1) State the developing history of leather industry in China.

(2) Give your description about current leather industry in Asia.

(3) Why does Europe stand out as the leading force in current international business circles related to leather?

(4) Give your description about leather industry in North America.

(5) Does Latin America have the potential for developing leather industry? Please give your explanation.

(6) Can camel hide be used as material for leather making? Please give your example.

(7) Why has leather industry been an important strategic sector for the economic and industrial development of many African countries?

Chapter 2 Basic knowledge of hides and skins

2.1 Kinds of hides and skins

2-1

The heart of the leather-making process is the raw material, hides, and skins. In principle, the hides and skins of nearly all higher animals can be used as the raw materials for leather making if they have a sufficiently strength fibre texture and are large enough to ensure economic production. Leather is mainly produced from the hides and skins of mammals which are bred for the production of meat and milk. Therefore the hides are mainly a side product of cattle breeding and the leather producing industry is a sort of disposal for valuable waste products. Hides and skins differ in their structure, depending upon the habits of life of the animals, season of year, age, sex, and breed. The younger the animal, the thinner and smaller the skin, the smoother and finer the grain structure, and the less likelihood of damage by diseases, insects, etc. The more natural the feeding and living conditions, the better the quality of the hide or skin. The female skin is usually finer-grained than the male and the fiber structure is looser, especially in the flanks, giving a somewhat softer leather.

In the terminology of the leather industry, the skins of large animals, such as full-grown cattle, cows, steers, buffaloes, yaks, and horses, are called hides. The term "*skin*" is used when referring to small animals, such as sheep, goats, pigs, and calves. The term "*hide*" is never applied to small animals. Based on the incomplete statistics, the proportion of different hides or skins are shown in Table 2.1.

Table 2.1 The different hides or skins and their proportion

The kind of hide or skin	The proportion	The kind of hide or skin	The proportion
Cattle hide (cattle hide, calf skins)	65%-70%	Pig skins	3%-5%
Sheep and lamb skins	10%-12%	Other types of skins	1%-2%
Goat and kid skins	8%-10%	Reptile and fish skins	below 1%

2.1.1 Cattle hide

Cattle hide is a general term of classification which includes Ox hide, bull hide, cow hide, steer hide, etc. The terms "bull", and "cow" was separately referred to full-grown male and female animals. "Ox" and "steer" denoted the castrated male. In addition, Cattle hide also includes the skins of beef cattle and dairy cattle.

Beef cattle are raised primarily for their meat, are kept confined and fed a diet of high-protein foods, while dairy cattle are raised mainly for their milk. The area of Cattle hide is large and its weight is generally from 15. 75 to 36kg (35-80lb). "Veal" denotes the skins of younger animals and its weight is from 6. 75 to 12. 25kg (15-25lb). "Calf" refers to young animals and the weight of its skin is from 2. 25 to 5. 4kg (5-12lb). For the best economic benefit or other purposes, young cattle were slaughtered and the calfskin was obtained. Compared with the characteristics of mature cattle hides, calfskins have the fineness of grain because the hair follicles are much smaller, collagen bundles are smaller and the follicles are much closer together. Calfskins are used for the manufacture of the finest leathers.

The quality of cattle hide is connected with the kind of cattle, its breeding methods and living conditions (Figure 2. 1). The best hides are from beef which are tough, uniform in thickness and have a "square" shape, i. e. less neck, leg, and belly. The skin of dairy cattle is generally loose-texture, less square in shape, and loose or thin in the belly. The quality of other cattle hides varies between that of the beef and dairy cattle. Cattle hide is used for furniture leather (including upholstery, sofa, and car-seat leather), shoe upper leather, bag leather, belt leather, etc.

(a) Ox (b) Bull

(c) Cow (d) Calf

Figure 2. 1 The picture of cattle

2. 1. 2 Buffalo hide

Buffalo is one kind of bovine (Figure 2. 2). Buffalo were mainly bred in China, India, Pakistan, and Indonesia. Based on the age, sex, and the living conditions, Buffalo hides have thicker, coarser, looser textures than that of Ox. In addition, they have badly wrinkled over the shoulder. They are preserved in the wet-salted state in China and exported in the dried, paste-dried, or vegetable-tanned state from India, Pakistan, and Indonesia.

2-2

Based on the characteristics of the grain and tissue, Buffalo hides are usually used for producing leather with natural texture (such as summer sleeping mat

(a) Buffalo (b) Yak

Figure 2. 2 The picture of buffalo and yak

leather, belt leather) or corrected leather (such as furniture leather, shoe upper leather).

2. 1. 3 Yak hide

Yak is also one kind of bovine. The yak lives in cold plateau areas in northwest and southwest China as well as Nepal, Pakistan, India, Russia, etc. The yak hide makes up 20% of total production of cattle hide. In China, yak hides are usually used for producing corrected leather such as shoe upper leather, and garment leather because of much defects and weak linkage between the grain layer and reticular layer.

2. 1. 4 Sheepskin

Sheep are raised primarily for their wool. The hair growth has been improved, resulting in very long fibers and fine texture through selective breeding. Since the sheep are protected primarily by the wool, the function of the skin is more to support the growth of the wool than to serve as a protective organ in itself. In a cross-section of sheepskin, a large number of fat cells are visible. The oil given off by these cells will lubricate the wool. Sheepskin is very open and porous and has very little structural fiber. This lack and the large concentration of glands in the area at the base of the hair root result in a physical weakness of the skin at this point.

Sheepskins have widely different characteristics depending on the condition of being bred. Generally, they are sub-classified as hair types and wooled types. In hot countries, especially those in the tropics, there are a number of varieties of sheep having hair instead of wool. Their skins have a fine, tight grain and flesh, and are very suitable for the production of the finest types of gloving and garment leather for both the grain and the suede finished. Wooled skins vary from about three to twelve square feet each, and the amount of wool can weigh more than twice as much as the skin proper. There are many distinct breeds including Australian Merino, British Breeds, New Zealand Breeds, etc. They have extremely long high-grade silky wool, but a weak, greasy pelt. They are shipped either in pickled or in a rough-tanned state, and are of lower sheepskin quality. Sheepskins are used for garment leather, boot leather, glove leather, etc.

A lamb is a young sheep whose age is below 1 year old. Because the fur of the lamb is fine, dense and the skin is fine, it can be made into double-face leather or leather and be used for carpets (Figure 2.3), thermal insulation products, and glove leather, etc.

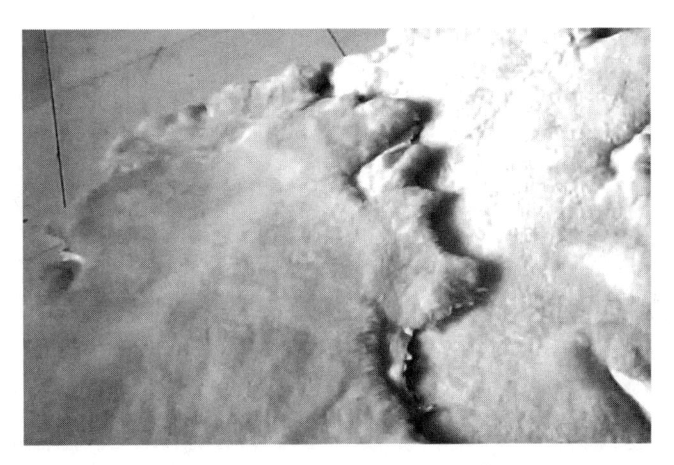

Figure 2.3 The double face leather made from lambskin

2.1.5 Goatskin

The goat is an ideal animal for tropical countries and localities where adequate grazing land for sheep or cattle is not available. The protection of the animal is partly from hair and partly from the fiber of the skin. Goatskins, as compared to sheepskins, have a very tight fiber structure, a very hard-wearing grain, and are easily recognized. The tight-natured fiber of goatskin allows its use in the more durable type of application in the manufacture of gloves and shoes. The raw skins are simply dried, wet-salted, dry-salted, and then baled for shipment to the tanneries (Figure 2.4).

The kid is a young goat whose age is below 1 year old. Because the lamb skin is fine, it is suitable for upscale leather.

(a) Sheep (b) Goat

Figure 2.4 The picture of sheep and goat

2.1.6 Pigskin

The structure of pigskin is in accordance with the habits of the animal. The domestic pig is protected by a layer of fat lying just below the surface of the skin. A pig has very little hair and its skin is a relatively tough tight-natured weave with a large quantity of stored food fat. The hair of the pig is relatively stiff, is set in small clumps, and the bottom of the hair follicle is very near the inside surface of the skin. Therefore, pigskins are essentially porous, having holes all the way through them due to the hair follicles. They are usually wet-salted or freshly put into leather production and are used for garment leather, glove leather, shoe upper leather, inner leather, etc (Figure 2. 5).

(a) Pig (b) Horse

Figure 2. 5 The picture of pig and horse

2. 1. 7 Horse hide

Compared horse hide with cattle hide and calf skin, it has a very peculiar fiber structure. Horse hide may be divided into two parts. The fore-part of the hide is known as the horse front, which is relatively light skin and in spite of a fairly heavy growth of hair, the texture of this area of the skin is not much different from some types of goatskins and is used for heavy gloving and shoe upper leather. The back portion of the hide, from the rump, contains a much thicker, less porous, and tougher area known as the crup/the horse butt shell which is a close network of fibers and has a very dense structure, is the source of the genuine cordovan leathers and is usually cut out and dressed separately. Today, with the development of leather technology, the whole piece of horse hide has been used for producing clothing leather. Horse hides usually are wet-salted or dry-salted. In China, most horse hides come from Inner Mongolia and northwest areas (Figure 2. 5).

2. 1. 8 Reptile skin

Reptiles are cold-blooded animals such as crocodiles, snakes, pythons, lizards, etc (Figure 2. 6). Their skins have no thermostatic function and therefore they are devoid of hair and fat glands. The scales are functionally and chemically related to the hair of warm-blooded animals. Histological investigation indicates the nature of the protection gained from the scales in this type of animal. In place of the hair and epidermis of mammalian skins, reptiles have a keratinous layer of scales, which is removed in the lime-yard. The weave of the fiber is different from those of mammalian skins, being much more horizontal and dense so that the skin tends to be very tough and thin, and less soft or supple than mammal skins.

(a) Crocodiles (b) Snake

(c) Python (d) Lizard

Figure 2.6 The picture of some reptiles

Crocodile skins are principally obtained from Africa, Central America, and Asia, and are usually wet-salted, well-preserved, and well-flayed. Depending upon the scale patterns, crocodile skins are divided into two types-large and small scale. The small scales are regarded as being the most desirable and are worth more than the others. African crocodile skins have a smaller scale pattern than Asia skins and have a longer body. Therefore the African crocodiles are worth more. The crocodile skins are used for belts, bags, etc.

Most snakes and lizards originate from India and Indonesia. Some of them come from Africa and South America. The skins are usually sun-dried, although shade drying is preferable and sun-dried material is often extremely difficult to soak back. Wet-salting and dry-salting methods are also used, and the salting is often done with mud which contains a high percentage of salt, rather than with pure salt. Lizards may be wet-salted as frequently as sun-dried. Pythons and other snakes are mainly air-dried. The snake skins and lizard skins are used for neckties, bags, shoe uppers, etc.

2.2 The structure of hides and skins

Most of all the raw hides and skins are from mammals. Therefore it is important to understand the nature of the hide and skin in order to rationalize the structure-function, structure-reactivity, and structure-property relationships during the process of leather manufacture.

2.2.1 The histological structure of hides and skins

2-3

Animal skins include hair-covering and skin. As a natural material, it is required to perform different functions and to deal with different stresses over its area. Consequently, skin is anisotropic and its structure and properties vary over its area. A typical shape of hide or skin and a division of the different parts are shown in Figure 2.7. The main parts contain the neck section, belly section, butt section, etc which vary widely in their thickness, fiber structure, and fiber density. The butt section is defined by the region up to halfway from the backbone to the belly edge and two-thirds of the way from the root of the tail to the neck edge whose fiber structure is relatively consistent and hence the physical properties of the skin or leather are consistent. The remaining regions to the side of the butts are the belly section and the remaining region beyond the butt towards the head is the neck section.

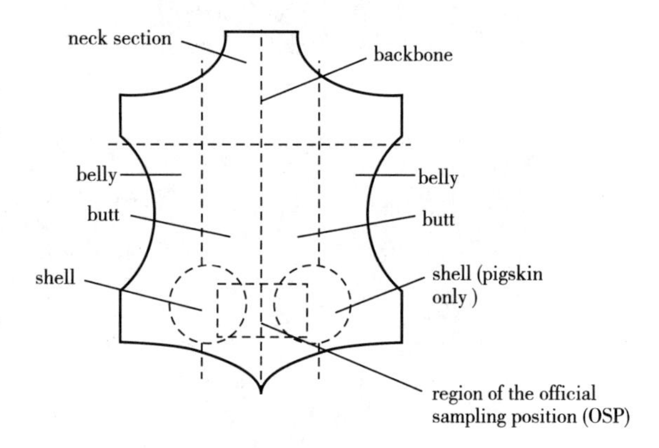

Figure 2.7 Different parts of hide or skin

Generally, the most valuable and uniform part of the hide or skin is the back and butt. The butt has a tight fiber structure, making the skin relatively firm and stiff. For pigskin, the butt section contains the shell area which is particularly hard. The belly section has a looser and less uniform fiber structure, varies considerably in thickness, and has an open structure so as to make them relatively weak. The neck section is the thickest part with a relatively open structure. It is an important aspect of the technology of leather making to try to make the non-uniform skin structure as uniform as possible in the final leather product. Therefore hides and skins are considered to be of good quality if they are relatively uniform in thickness and fiber structure. Poor quality hides and skins are those with a loose, uneven fiber structure and substantial differences in thickness between the butt and belly sections.

As mentioned previously, skin is anisotropic and the anisotropy varies over the whole skin. In order to make the best comparisons of leather properties, it is necessary to adopt *the idea of an Official Sampling Position*. Consequently sampling for physical testing is routinely done in two ways, parallel and perpendicular to the backbone. The region of the backbone extends from the tail to neck and a few centimeters on either side of that line. It is thicker than the butt and must be shaved to the thickness of the butt region.

The histological structure of the hides or skins is more or less the same because the main purpose of the hides or skins of all living animals is to protect the animal from injury and the effects of the weather, to regulate their body temperature, etc. The cross-section structure of the hide or skin is shown in Figure 2.8 which represents the generic structure of most types of animal skin including human, cattle, pig, goat, and sheep skins. There are minor differences among the different types of animal skin. From the leather point of view, they consist of three layers, i.e. epidermis, corium or cutis, and subcutaneous tissues (sub-cutis).

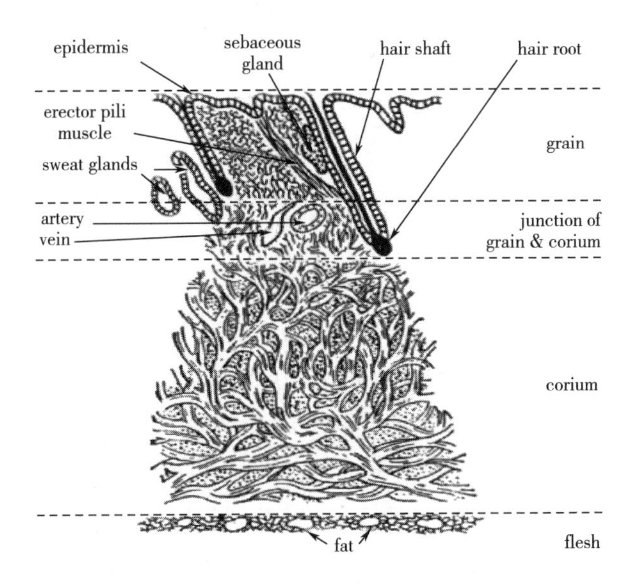

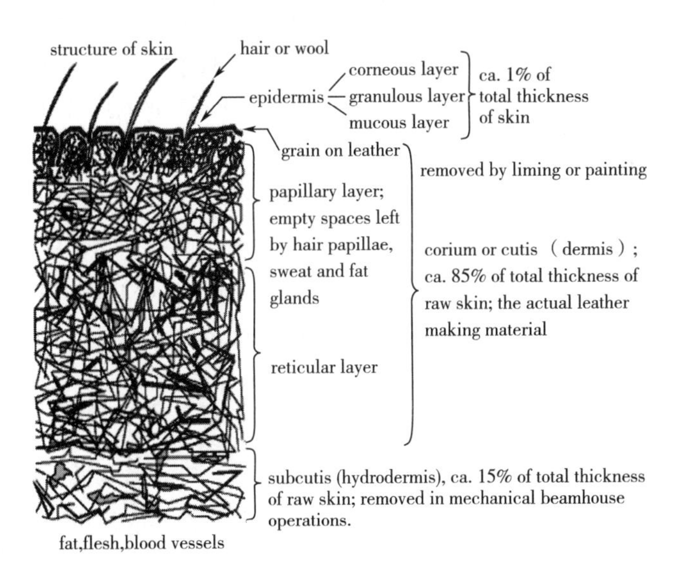

Figure 2.8 Cross-section of hide or skin

2.2.1.1 Epidermis

The epidermis is the outermost layer of the raw skin, is the barrier between the animal and its environment, and is a protective, hard-wearing layer composed of keratinous cells. The epidermis is quite inert chemically, but is easily decomposed or disintegrated by alkalis, such as caustic soda, lime, and especially sodium sulfide or hydrosulfide, and falls off the skin as scurf or dandruff. In addition, the cells from the underside of the epidermis are forming and pushing outward, giving constant new growth to protect the outer layer of the skin because they consist of soft, jelly-like living cells which have little resistance and are readily attacked by bacterial action or enzymes.

In the beam-house processing of leather manufacture, the epidermis is removed when hair or wool is removed from the skin, particularly by chemical dissolving techniques.

2.2.1.2 Corium

The corium is the main part of the hide or skin which has an obviously fibrous structure. The fibrous structure varies through the cross-section of hide or skin. The fibers increase in size, reaching a maximum fiber diameter in the center of the corium and then decreasing a little as they approach the next lower layer.

The network of fibers is very intimately woven and joined together. It consists of fibers dividing and recombining with other fibers so as to make the corium strong and able to resist stress placed on it. The corium is subdivided into a grain layer and a reticular layer. The grain layer is the uppermost layer in unhaired or de-wooled pelt whose collagen fibers become very thin, tightly woven and so interlaced that there are no loose ends on the surface beneath the epidermis. When the epidermis is carefully removed, a smooth layer is revealed known as the hyaline layer which gives the characteristic grain surface of leather. The fibers of the reticular layer are coarser and stronger than those of the grain layer. The fiber structure here is fairly dense, heavy, and tangled.

Another important feature of thereticular layer is the "angle of weave" which affects the physical properties of the leather. The magnitude of the angel can provide useful informationable regarding the process history of the pelt. That is, the angle varies in different animals, but to some extent, the angle can be altered during the leather manufacturing process. The average angle of the weave is about 45℃. A lower value indicates greater depletion of the corium and a higher value indicates the degree of swelling. In addition, the fiber structure indicates the properties of the finished leather. The leather is firm, hard, and little stretch if the fibers are more upright, and tightly-woven. Whereas the leather is softer, weaker and less elastic if the fibers are more horizontal and loosely-woven or a low angle of the weave.

2.2.1.3 Subcutaneous tissues

The subcutaneous tissue layer (sub-cutis) of hide or skin is close to the "meat" or "flesh", where the fibers are loosely woven and have a more horizontal angle of weave, and fatty (adipose) tissue may also be present. This variation of the thickness of the subcutaneous tissue layer is affected by many factors such as age, sex, breed, health and nourishment of the animal, etc. The layer is usually removed by the fleshing machine during the process of leather manufacture.

2.2.2 The chemical components of hides and skins

Fresh hides or skins are made up of water, protein, fatty materials, and some mineral salts, carbohydrates, etc. The approximate proportion of different compositions of hides or skins is indicated in Table 2. 2. Of these, the most important for leather manufacture is protein.

Table 2. 2 Composition of hide and skin

Composition	Water	Protein	Fats	Mineral salts	Carbohydrates
Percentage/%	60-65	30-35	2. 5-3. 0	0. 3-0. 5	<2

Types of this protein have structural proteins and non-structural proteins. Structural proteins consist of collagen, keratin, elastin, etc. During the process of leather manufacture, the most important protein are collagen which can be converted into leather by tanning. Keratin is the chief constituent of hair, wool, and the epidermal structure. Non-structural proteins consist of albumens globulins, mucins mucoids. The approximate proportion of different proteins of hides or skins is indicated in Table 2. 3. The proportion of the components of hides or skins is associated with the kind, sex, age, and living condition of the animals.

Table 2. 3 Types of hide's protein and its proportion

Type of protein	Structural proteins		Non-structural proteins		
Name of protein	Collagen	Keratin	Elastin	Albumens globulins	Mucins mucoids
Approximate Percentage/%	29	2	0. 3	1	0. 7

2. 2. 2. 1 Structural proteins in hide and skin

It can be seen from Table 2. 3 that structural proteins consist of collagen, elastin, keratin, etc. Based on their structure, they have different properties respectively.

- Collagen

2-4

Collagen is the main content of the corium of hide or skin. It forms a three-dimensional, irregular network of fibers, ramified and interwoven in all directions without beginning or end. Based on the visible through a microscope, each of the collagen fibers consists of 30-300 "elementary fibers" (diameter approx. 0. 005mm), which in turn are each made up of 200-1000 "fibrils" (diameter approx. 0. 0001mm). Each fibril contains 700-800 collagen molecules. The collagen molecule consists of three peptide chains which are composed of 1, 052 amino acid radicals. The collagen is twisted with each other in the form of a triple helix, which is approx. 0. 003mm long, has a diameter of approx. 14×10^{-7} mm and a molecular weight of approx. 300, 000. The triple helix, simulated diagram of collagen is shown in Figure 2. 9 and Figure 2. 10. The triple helices are linked through hydrogen bonding which is an important feature of the structural or chemical stabilization of native collagen.

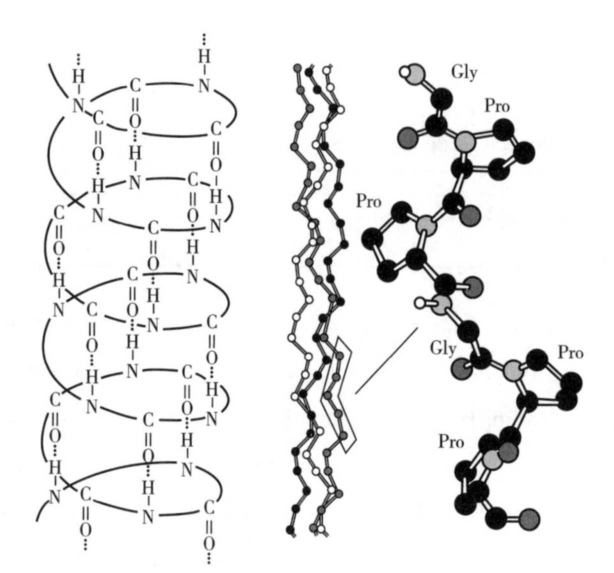

Figure 2. 9　The triple helix of collagen fiber

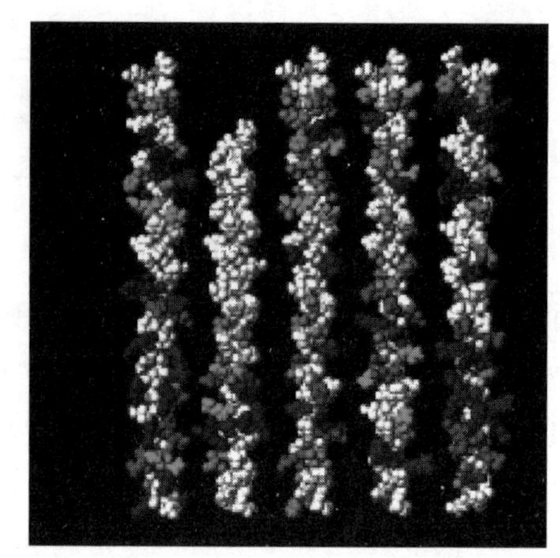

Figure 2. 10　The simulated diagram of collagen

The triple helices are bound together in bundles called fibrils. Fibrils are arranged in the form of fibril bundles in which the fibril bundle is a sub-structure or the constructing units of fibers. With respect to the opening up of the fiber structure, this level of the hierarchy of collagen structure is important in preparation for the tanning process. At the junctions between fibril bundles, splitting open the fiber structure at this level is important for creating softness and strength in the finished leather, particularly in regard to the deposition of tanning agents, and lubricating agents.

The collagen contains acid and basic amino acids (Figure 2. 11), as well as a relatively high percentage of glycine, proline, and hydroxyproline. Collagen fiber is insoluble, unless subject to heat or bacterial degradation and can't be broken down in a neutral solution. But it swells by absorbing water in acid or basic solution. Collagen fiber may be hydrolyzed to dissolve in water when a very strong acid or alkali is added under high temperature.

(a) Clycine

(b) Proline

(c) Hydroxyproline

Figure 2. 11　The structure of some amino acid

In addition, collagen can react with chromium tanning agents, aluminum tanning agents, aldehyde, vegetable tannin, as well as other cross-linking agents so that structural stability can be greatly enhanced and high hydrothermal stability of collagen fibers is obtained. This is the basis of the common tanning process that makes the skin into leather. Figure 2. 12 shows the expected effect of the chrome tanning process based on hide powder.

- Elastin

Elastin is the second most important protein in the skin after collagen. Because elastin controls the elasticity of the grain layer, it greatly gives the physical properties of hide or skin and leather. The

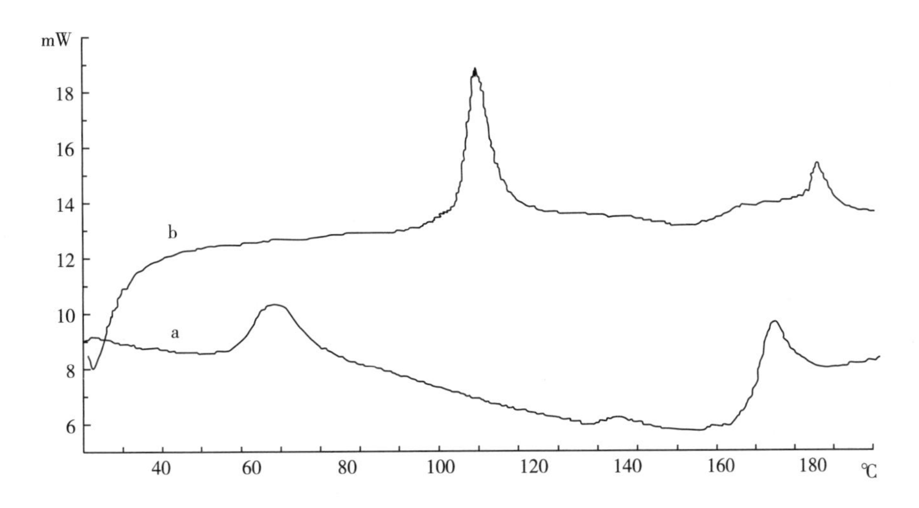

Figure 2. 12 DSC thermograms of bovine hide collagen

a—Untanned hide powder b—Chromium (Ⅲ) tanned hide powder

material of the grain is weak and can not stretch to accommodate stresses in the skin when, for example, a joint is flexed. Hence it adopts a convoluted, rippled form, that can flatten as the corium stretches. The mechanism by which it returns to its convoluted state when stress is removed through the action of elastin fibers. The elastin fibers are centered on the follicles, with coarse fibers running parallel to the hide or skin surface and fine fibers running at right angles to the skin surface.

The content of residues of amino acid per 1,000 of elastin and collagen is shown in Table 2.4. Compared with collagen, elastin has similar glycine and proline content which indicates the structure of elastin could be helical. In addition, elastin includes more apolar amino acid, little acid, and basic amino acid, and less hydroxyproline so that the protein is more hydrophobic and has less hydrogen bond than collagen.

Table 2. 4 Comparison of elastin and collagen: residues per 1,000

Type of amino acid	Elastin	Collagen
Glycine	355	330
Apolar	431	170
Acidic	14	120
Basic	10	196
Hydroxy	20	57
Proline	12	126
Hydroxyproline	23	93

Different from collagen, elastin is remarkably inert to chemical reaction. If hide or skin is placed in a solution of 0.1mol/L hydrochloric acid and is heated, the hide structure will be destroyed, but a network of elastin will be intact. The elastin network doesn't have sufficient change and isn't solubilized under the conditions of strong acid. Elastin also doesn't play a role in leather and is hydrolyzed in the bating process.

- Keratin

The main component of hair, epidermis, and hair sheath is keratin which is made up of amino acids containing both acid and basic groups (Figure 2.13). Different from collagen, keratin has relatively large quantities of the sulfur bearing amino acids, e. g. cystine and methionine which are more resistant to acid than collagen. Based on the different content of cystine, keratin in different tissues may show different chemical activity. For instance, the keratin contains quite little of cysteine. i. e. in the hair root which has not yet formed complete cystine cross-linkages. Therefore the keratin has not attained an inert structure and is easily hydrolyzed by alkali. However, the keratin contains relatively large quantities of cystine in the hair shaft, it is most inert and little subject to alkaline hydrolysis due to the scale on the surface.

(a) Cystine (b) Methionine

Figure 2.13 The structure of some amino acid

2.2.2.2 Non-structural Protein in hide and skin

It can be seen from Table 2.3, that non-structural proteins consist of albumens, globulins, mucins mucoids, etc. Based on their structure, they have different properties respectively and should be removed in the process of leather-making.

2-5

- Albumen

Albumen is one of the soluble proteins characterized by having a high percentage of acid and basic amino acids. The protein is highly ionized over a large portion of the pH range, and there are a large number of charges per unit weight. These charged sections of the protein have some electrostatic attraction for one another giving the molecule a tendency to fold back upon itself. This forms what are called molecular globules; such proteins are known as globular proteins. If a high percentage of salt (over 10%) is added to a solution of a globular protein, say albumin, the salt may act as an electro-chemical bridge between adjacent protein molecules and result in the precipitation of some of the globular proteins. On the other hand, a lower concentration of salt, in the neighborhood of 5%-10% or lower, depending upon the protein, may aid in the solubilization of the protein. For this reason, in the salt curing process, some of the globular proteins are leached out by the salt.

If a solution of albumin is raised in temperature to its critical point, the molecular agitation may become sufficient to overcome the intra-molecular electrostatic attraction and allow the adjacent molecules to come into contact. At this point cross-linking between molecules may occur and essentially the whole protein solution may become one large molecule through mutual inter-molecular electrostatic attraction. This is the reaction that is observed in the insolubilization of albumin in the boiling of an egg.

- Globulin, mucins, and mucoids

Globulin, very similar to albumen, is a non-fibrous protein. Its molecule is also globular. It is soluble in a dilute solution of salts, alkalis, and acids. It is often removed in the liming process, etc. Mucins and mucoids, like albumen and globulin, are interfibrillary proteins. They are soluble in dilute salt solution and can be hydrolyzed by pancreatin.

- Hyaluronic acid

In addition, the skin contains hyaluronic acid (HA) and dermatan sulfate (DS). The structure of hyaluronic acid (HA) is a chain of alternating D-glucuronic acid and D-*N*-acetyl-glucosamine, joined by β-1,3 glycosidic links, with a molecular weight of the order of 10^6, corresponding to 3-19000 disaccharide units. They pack or gather between the fibers or on the fibers and give compressive resistance to the skin or hide.

Hyaluronic acid is very gelatinous, and it is difficult and important to get out during the soaking process. It can be considered as a polymeric chain of aliphatic carboxylate groups with pK_a about 4.0 which is ionised at pH 7.4 so that it can react with chromium tanning agent or other miner tanning agents and obtain stiff leather. Therefore hyaluronic acid (HA) should be removed prior to tanning by alkali treatment with $Ca(OH)_2$ in the liming process.

- Dermatan sulphate

Dermatan sulphate proteoglycan has a protein core and 2-3 side chains of 35-90 repeating units containing one molecule of L-iduronic acid and one molecule of N-acetyl-D-galactosamine-4-sulfate whose molecular weight is about 10^5 (60% is protein and 40% is polysaccharide). Dermatan sulphate proteoglycan is bound to the surface of the fibrils in the hierarchy of collagen by protein links and electrostatic interactions.

Dermatan sulphate proteoglycan's works are similar to hyaluronic acid in some ways. During the liming process, the strong alkali attacks the covalent bond between the polysaccharide side chains and the protein core and they essentially fall off the fibers so as to achieve the opening up and the separation of the fibrils.

2.2.3 Hair or wool and other components of hide or skin

Besides the structure of cross-section of hide or skin, other components include hair or wool, follicles, erector pill muscle, sweat glands, veins, and arteries, collage, elastin, etc.

2.2.3.1 Hair or wool and follicles

In general, all mammal hides or skins have hair or wool, but they vary in size, the shape, and thickness from the hide of the elephant to the skin of the rabbit. Some animals have little hair or wool and a thick epidermal layer, while others such as the sheep, have a heavy fleece and a thin epidermis. There are great variations even in one skin. The hair is usually denser and coarse on the back than on the belly, and the skin and epidermis thicker and tougher on the

2-6

more exposed parts than on the sheltered places in the limb joints. The variation is affected by age, sex, breed, health, and nourishment of the animals.

The hair or wool is situated in follicles and consists of the shaft, root, and bulb. The cross-section and its follicles of hairs are shown in Figure 2. 14. The part of hair located outside of follicles is shaft, and the part of hair located inside of follicles is root and bulb. Bulb-containing the papilla is the structure from where the hair grows and this is the bottom part of the hair and gives the nutrition for the growth of the hair.

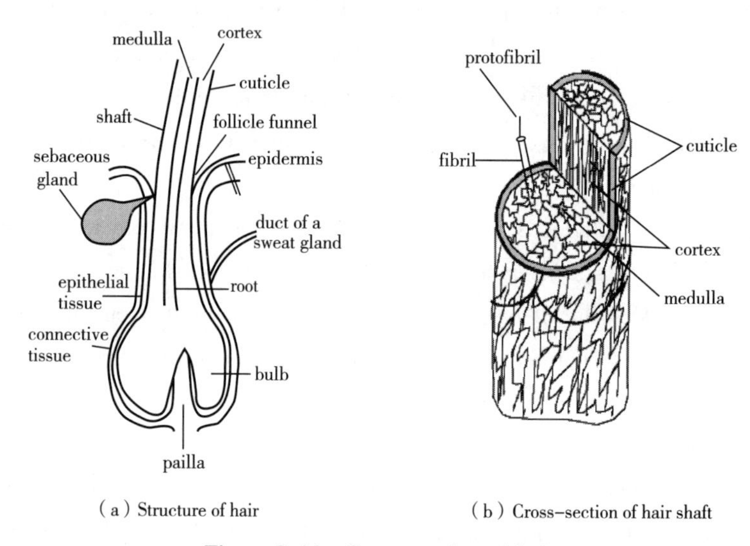

(a) Structure of hair (b) Cross-section of hair shaft

Figure 2. 14 Cross-section of hair

The hair follicles provide the grain pattern which is so much a feature of the value of the leather. The follicle angle reflects the angle of the weave of the corium. But the corium angle increases and decreases if the corium angle decreases so that the angle of the weave is a reflection of change in area. It is important that the follicle angle should be low in the final leather because it determines the fineness of the grain pattern. The higher the angle of the follicle, the more the mouth of the follicle appears open and obtains coarse, open, or grinning grain which is undesirable. In terms of both regularity and distribution, the grain pattern is uniform in most animals. For instance, the arc pattern of pig bristles and plumes or quill of ostrich.

In addition, the hairs conclude the cuticle, cortex, and medulla from the viewport of a cross-section of hair shaft.

Chemically the hairs consist of the keratin in the form of cells, dead and resistant at the tip and easily destroyed at the root. During the process of leather-making, the hair or wool is commonly removed by in various ways such as hair-destroying methods with sodium sulfide and hair-saving methods with enzymes. But if the hide or skin will be made into double-face leather for coats or rugs, the hair or wool will be left, e. g. wool of sheepskins.

2. 2. 3. 2 Other components of hide or skin

Besides the protein and hair or wool, etc in hide or skin, it also includes erector pili muscle, sweat glands, veins, and arteries, etc.

- Erector pili muscle

The erector pili muscle is the important part of the hide or skin which can help animals resist the cold by raising the angle of the follicle, to trap a layer of air within the hairs and by contracting the muscle pulls across the upper part of the skin. The phenomenon called "goose flesh" or "goose bumps" will be created for humans or living animals. Therefore, the preliminary soaking step may result in coarse grain under cold conditions. The muscle is degraded and hydrolyzed during the liming and bating process so that its action is no longer available to influence follicle angle and the quality of the grain.

- Sweat glands

As shown in Figure 2.8, sweat glands are an important part of living animals which can help them to adjust the temperature of the body. But sweat glands do not play any role in leather processing. Their components are degraded in the early alkali-based process and their removal does not influence leather properties or quality.

- Veins and arteries

The blood vessels are a series of vessels for the flowing of the blood which can provide the nutrition for living animals. The veinous system of the hide or skin is located in the junction between grain and corium, where the main vessels run parallel to the skin surface, fed by smaller vessels running vertically down into the body of the animal. The veinous system is largely composed of elastin because of the requirements of flexing as the heart pumps blood around it.

Veinous systems also do not play any role in leather processing. Veins and Arteries will not affect the quality of the leather if efficient bleeding is achieved and to empty the blood vessels at slaughter. However, if it is not achieved or the blood vessels are degraded during processing, a defect called "veininess" may be observed. The unmistakable appearance is a reflection of the blood vessel network on the split (corium) surface. The pattern occasionally may be observed on the grain surface in severe cases. Therefore the appearance of a pattern of blood vessels may arise from two different mechanisms. Inadequate bleeding in the abattoir leaves the vessels expanded because they still contain blood. Although the blood is degraded and removed during the liming process, the veinous vessels do not collapse completely and create voids in the hide or skin. The blood vessels will be dissolved by the elastolytic enzymes in processing and voids are left in the pelt cross section.

2.3 Curing and preservation of hides and skins

Based on the components of hides and skins, they include lots of nutrients so as to be invaded by microorganisms. Once the hide or skin has been removed from the animal, it is subject to degradation from cellular enzymes from within the skin and from enzymes produced by bacteria from the environment. During putrefaction bacteria penetrate the hides and skins, producing enzymes which can degrade skin components. This can result in an off-smell, rancid fats and breakdown of the epidermis. Therefore raw stock without

2-7

preservation will rapidly degrade. Hides and skins have to be cured or preserved effectively to prevent putrefaction and deterioration.

The function of curing and preserving rawstock is to stop all types of degradative enzyme activity. Curing or preservation treatments have to be carried out immediately. Curing is one type of preservation method where rawstock is treated with salt. Preservation is normally achieved by inhibiting the growth and action of bacteria (or fungi) and hence preventing the production of degradative enzymes. Preservation is accomplished either by destroying active bacteria, by preventing bacterial activity, or by preventing bacterial contamination. Ideally, any treatment applied to the pelt should be reversible, without altering the properties of the pelt.

Various methods can be applied to the cure and preservation of raw-stock. The most common method of preservation is treatment with common salt. In addition, drying, chilling, freezing, irradiation, treatment with biocides can also be employed.

2.3.1 Drying

2-8

The living of bacteria requires a certain amount of free water or moisture. If the water is removed and the hide contains only 10% - 14% moisture by the drying process, activities of bacteria cease and some types of bacteria are killed. Thus they can remain a long time until there is enough water for them to be active again.

Drying is practiced in some countries or districts with hot, dry climates such as India, Africa, South America, North China. The operation is that the hides or skins may be dried by simply spreading them out on the ground or hanging or laying them over poles, ropes, or wire in the sun (hanging drying). The former may be dangerous owing to poor ventilation of the ground side and the high temperature of the exposed side. The latter gives better ventilation and quicker drying, but may result in heat damage and pole or rope marks, showing as hard creases down the hide. In addition, frame drying is adopted sometimes. Hides are dried by loosely straining them out on frames so that they don't receive the direct rays of the mid-day sun. This gives less danger of heat damage and a better, flatter shape. Frame dried hides are of better shape, more uniformly dried and less liable to putrefaction or heat blisters than the ground dried and hung dried.

Velocity of drying is very important for the quality of the dried hide or skin, especially thick hide. Putrefaction may occur before the moisture content is low enough to stop bacterial action if drying is too slow such as in a cold, wet climate. But if drying is too fast and the temperature is too high, part of the wet skin will start to gelatinize to a glue-like material which makes the skin hard, brittle and prevents drying of the inner layers. The fault is difficult to be distinguished in dried hides until they are soaked back in water. Owing to the gelatinized part of the skin dissolving in water, holes appear or the smooth grain is lost or has a blistered appearance.

Dried hides and skins are susceptible to be attacked by insects such as beetles, larvae, and maggots. A common method of prevention is either to dip or to spray the hides with a solution of white arsenic and caustic soda. Other methods use proprietary insecticides such as sprays, dips, or dusting powders.

2.3.2 Salting

Most putrefying bacteria and their enzymes require a wet substrate for active life. Therefore reducing free water is an obvious method of stopping putrefaction. The most common and ancient method of preservation is by the application of common salt. The mechanism is that the osmotic effect of a high concentration of salt on the outside of the pelt, with the surface acting as a semi-permeable membrane, causes the water in the pelt to migrate to the outside and the salt to migrate to the inside. The removal of water reduces the viability of bacterial activity. The high concentration of salt within the pelt creates an osmotic effect on the cell wall of the bacteria, reducing its viability further.

The common practice of salting is to distribute salt over static pelt, or spread salt by kicking a pile across the pelt on the ground is the common practice. It is necessary to achieve at least 90% saturation (i. e. >32% w/v in the moisture in the pelt, about 14% on hide weight) within the pelt so as to ensure no bacterial activity. Therefore the methods of the curing and preservation of salting include stack salting and brine salting.

The way of salting is applied depends on the scale of abattoirs. In large abattoirs or curing plants, the hides and skins are brined in drums. However, stack salting is the most appropriate method of cure in medium-sized abattoirs handling 400-500 cattle per day.

2.3.2.1 Stack salting

Stack salting is the most common method of wet salting. The freshly flayed hides or skins are spread out on the ground, flesh side up, then salt is sprinkled or spread out over the whole of the surface. Alternatively, the salt (sodium chloride) may be applied automatically, using a conveyor and automatic dosing. Hides are treated with about 25%-30% salt, typically about 10kg. Skins are treated with up to 40%-50% of salt, typically about 1 kg. A second hide is placed on the first one and also sprinkled with salt. The salted pelts are stacked in piles 1.5-2.5 meters high, often folded into quarters, with additional salt in between each layer and on the top of the pile. The piling helps to squeeze brine out of the pelts, leaving 40% - 50% moisture. The distribution of salt is close to equilibrium through the cross section after 24 hours and equilibrium is effectively reached in 48 hours.

2.3.2.2 Brining

An alternative to wet salting is to apply the salt in the form of a solution. Brining is a more efficient method for the curing or preservation of hides or skins. The hides are cleaned by hosing with water and are then tumbled in a paddle with saturated brine (36% w/v at 15℃) for 12-14 hours which gives very good and uniform salt penetration for heavy hides. The hides are then drained and piled. If prolonged storage or transportation is needed, the hides may have additional salt, with or without additive, applied to the flesh side after draining. Brined hides contain 85% saturated salt solution and are wetter than wet salted hides.

As the brine liquor may be contaminated with halophilic and halotolerant bacteria, the purity and strength of the brine liquor must be checked before it is reused. These bacteria are able to grow or live on high concentrations of salt and often give rise to red or colored patches on the flesh side, called

"Red Heat". Their presence suggests that some putrefaction has occurred which will weaken or damage the skin. Effective control against red heat is to apply biocides mixed into salt. The biocides in common use are boric acid or sodium metabisulphite mixed at the level of 2% on the salt weight.

However, the effect of salt doesn't act as a bactericide and can't kill the bacteria. Collagenolytic activity is still rapid at 7% salt on the weight of the skin, but low at 10%. The bacteria is still function. This means that the bacteria are still alive, still viable, and can be reactivated when the pelt is rehydrated. If the salting is done well, the rawstock can be kept for several months.

2.3.3 Chilling and freezing

Rawstock can be preserved for a short period by storing it at a temperature that slows bacterial activity. Therefore chilling and freezing are employed universally for the preservation of hides and skins.

If the hides are stored at 2 - 3℃, they will remain fresh for up to three weeks without the application of a biocide. The moist air should be used during the chilling, otherwise the rawstock will dry. The chilling operation requires a 3-meter long cooling tunnel through which hides would pass in 20 minutes. The hides collected within 4 hours of flay are suspended over poles, fleshed out, and placed on wheeled trolleys. The trolleys are loaded into a chilling room large enough to hold the day's supply of hides and they are left to chill under conditions which reduce the hide temperature to 2℃ in 1 hour but the hides remain overnight. The following morning the hides are folded hair out and piled into cage pallets holding 20-25 hides for storage at 1℃. The refrigeration is sufficient to maintain a temperature of 1℃ in the area. The insulating properties of the pallets of chilled hides can be safely transported to tanneries during summer with a delay of 36 hours before going into the process.

Chilled hides can be frozen and held in cold storage for long-term preservation. The cage pallets holding 20-25 chilled hides are frozen at-10℃. The time required to freeze the center of the pallet is 3 days. These pallets are held in cold storage for 6 months. The hides are thawed within 24 hours by immersion of the entire pallet in water and the water circulates. Satisfactory leather are produced from these hides.

It is necessary to freeze the rawstock or use other methods of long-term preservation (such as salting or drying), this will vary depending on the geographical region and type of rawstock.

Energy costs and the capital involved in holding stocks of hides in cold storage prevent the adoption of cold storage commercially at the present time (Table 2.5).

Table 2.5 Compared the time of preservation of different chilling degree

<table>
<tr><td>Ice</td><td rowspan="2">Short-term preservation (up to 3 weeks)</td></tr>
<tr><td>Chilling with refrigerated air</td></tr>
<tr><td>Freezing</td><td>Long-term preservation (over 3 weeks)</td></tr>
</table>

2.3.4 Use of biocides

"Biocide" is the general term which refers to both bactericides (which are effective against bacteria) and fungicides (which are effective against fungi), as shown in Figure 2.15. They include phenols, 2-

(cyanomethylthio)benzothiazole (TCMTB), methylene bis (thiocyanate) (MBT), 1,2-benzisothiazolin-3-one (BIT), biguanide, sodium chlorite, boric acid and 1,3-dihydroxy-2-bromo-2-nitropropane (bronopol). Their mechanisms include inhibition of bacterial enzyme systems or by reacting with the proteins of the microorganisms.

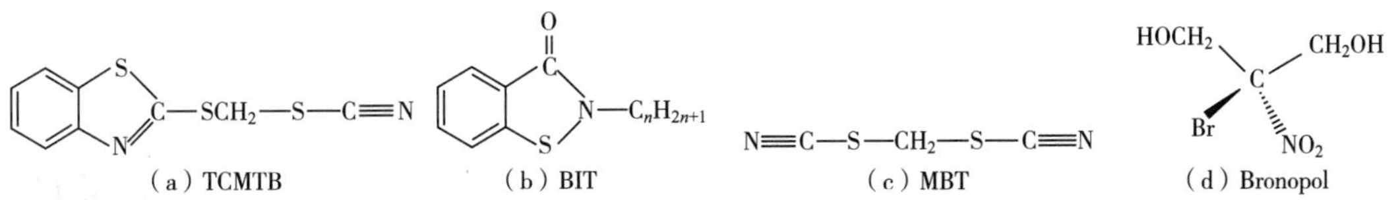

(a) TCMTB (b) BIT (c) MBT (d) Bronopol

Figure 2.15 Structures of some biocides applied in the preservation of hide or skin

Short-term preservation can be achieved by spaying hide on both sides with a solution of a bactericide so that it can be used as an alternative to salt for "short-term preservation" of rawstock for 3-21 days. The length of storage time depends on the agent and the concentration used. Longer storage is generally not possible due to re-contamination of the skin or breakdown of the biocide. Biocides are no cheaper than salt to use and have a number of other disadvantages: the health and safety implications of using the products, environmental problems, and potential resistance problems.

2.3.5 Radiation curing

Radiation curing is an important way in the modern preservation of rawstock. Up to now, gamma radiation and electron beam radiation have been investigated in the preservation of hide or skin.

Gamma radiation: The hide or skin can be protected for 6 weeks when gamma radiation provides complete sterilization at a dose of just over 20kGy (Gy = gray). The radiation is capable of damaging the pelt, reducing the tear strength, but the damage was not apparent below 30kGy.

Electron beam radiation: The procedure was to rinse hides with biocide, seal them in polythene bags, and irradiate with 1.4Mrad at 10MeV, which provided protection for 3 weeks.

These technologies can keep the hides sterile and avoid recontamination in spite of the cost of the plant for treatment.

Questions

(1) What are the sorts of animal hides or skins and their characteristics in the leather industry?

(2) What are the histological structure and components of animal hides or skins?

(3) What portion of animal hides or skins can be changed into leather in terms of the histological structure and components of animal hides or skins?

(4) What's the structure of the hair or wool?

(5) Why is the animal hide or skin easy to be invaded and degraded by microorganisms?

(6) Succinctly describe the methods of curing and preservation of hide or skin and related advantages, and disadvantages.

Chapter 3 Traditional technology of leather manufacture

3-1

Before hides or skins are processed, trimming and sorting are necessary to ensure relative uniformity during the production. In addition, the traditional production of leather manufacture can be looked upon as taking place in three steps. The first step is wash hides and skins from dirty such as manure, dung, blood and other residuals from animal or during the stockage, to remove the unwanted components such as hair, fats, etc, and open up the collagen fiber bundles, leaving a network of fibers of hide protein. The second step is to make this network react with tanning materials to produce a stabilized fiber structure. The production of wet-blue or wet white involves the first step and the second step. The third step is to build onto the tanned fiber characteristics of fullness, color, softness and lubrication, and to finish the fiber surface, to produce a useful, fashion able product so as to achieve functional and fashionable leather articles.

3.1 Trimming and sorting

The object of trimming and sorting is the preparation of skins for processing. The first few steps in getting skins and hides ready for processing take place in the tannery's hide house. The hide house is a large storage area that is kept cool and well-ventilated. It is here that the tanner receives and stores the hides and skins. These hides are all in a cured state.

3-2

The initial tannery operation consists of trimming off the heads, long shanks, and other perimeter areas. These offal areas generally do not make good leather and, if left on, would interfere with much of the tannery equipment through which the hide will be processed.

As a further aid to easier handling, the hides are cut lengthwise along the backbone head to tail to make two sides. This is the origin of the term side leather, or in other words, leather that is processed as two sides rather than one whole hide. The term bears no connection to the type of tannage, color, etc, of the resulting leather.

Traditionally, tanning is a so-called batch process. The next step, therefore, calls for gathering a number of sides, usually totaling 5,000 to 10,000 pounds, to form a pack. Each pack is properly identified as to size, weight, type of skin, and any other information that will be helpful to the later processing.

3.2 Beam-house processing

The first step is called beam-house processes because they traditionally included operations that were conducted over a wooden beam, as shown in Figure 3.1. This working of the skins on the beam is of ancient origin and is still in use today. Even in the most modern and sophisticated facilities, some hand operations on the beam are needed occasionally for quality improvement. Prior to the introduction of machinery, hides or skins would be draped over an angled wooden beam, so that they could be hand fleshed, hand dehaired (scudded), and hand shaved to thickness. The purpose of the beamhouse is to purify the pelt or "opening up" the pelt structure and prepare the pelt for tanning by the removal of non-collagenous components of skin such as the hyaluronic acid and other glycosaminoglycans, the non-structural proteins, the fats, etc, and splitting, separating the fiber structure at the level of the fibril bundles.

Figure 3.1 The tanner is removing the flesh of the hide with a knife on the beam

In the tannery, the term "beamhouse" refers to the processes between the removal of the skins or hides from storage and their preparation for tanning which includes soaking, trimming, fleshing, unharing, liming, scudding, deliming, bating, degreasing, pickling, depickling process, etc. Each step is important so that subsequent process steps can never compensate for or overcome deficiencies in any given prior process step-at least not without compromising leather quality.

3.2.1 Soaking

Soaking is the first process applied to the rawstock whose dominant function is making the preserved hide or skin (dried or dry salted hides, wet salted stock) back to the fresh state. The function of rehydration is to fill up the fiber structure with water, ensuring that all elements of the hierarchy of structure are wetted to equilibrium and beyond, with the purpose of facilitating the movement of dissolved chemical reagents through the pelt cross-section. Fresh hide requires filling with water, which is a faster reaction than any rehydrating process. In addition, soaking can remove salt, non-structural proteins, dung, and hyaluronic acid and clean the pelt.

Soaking is generally accomplished by placing the skins in a drum or paddle with water which may contain auxiliaries such as detergents, soaking enzymes, biocides, etc. Wet salted hides may be soaked in a drum or paddle full of water for 8-20 hours. The amount of water used is 2-5 times the weight of the hides. The water will dissolve the curing salts and decrease the concentration of salt around the fibers of the skin. This removal of the salt outside of the fibers causes an osmotic take-up of water into the hide fibers, and the skin will become rehydrated. Therefore the hide or skin increases in weight owing to the absorption of water. Dry hides or skins require longer periods of soaking for complete rehydration. Simultaneously some blood, urine, and body fluids from both surfaces and within the pelt are removed so that the hides become cleaner and softer in the soaking process.

The soaking operation includes pre-soaking and main soaking processing. There is no strict boundary between the pre-soaking and main soaking processing. Pre-soaking (or dirty soaking) is necessary when tanneries flesh the raw hides and the skins (green fleshing or pre-fleshing), which can remove part of salt, dirties and soften the pelt. The time of pre-soaking process is depending on the state of hides or skins. Aim of main soaking processing is to accelerate the re-hydration of hides or skins. Tanners can check satisfactory soaking by observing a uniform color in the cross section and using liquid universal pH indicator, generally green as pH at the end of the soaking is 8.5-9.5, which is shown in Figure 3.2. Also, a good indicator of the soaking for salted hides and skins is the salinity expressed in Baumé which is stable and does not increase. This means that the salt concentration is balanced in the leather and the float. Satisfactory soaking can be also judged by feel, cleanliness and the absence of salt.

Tanners use alkalis in the soaking. Sodium carbonate (soda ash) or magnesium oxide are typically used. Sodium carbonate, despite its low costs, raises the pH very quickly and can mark grown wrinkles. Magnesium oxide is more uniform but pH cannot get more than 8.5.

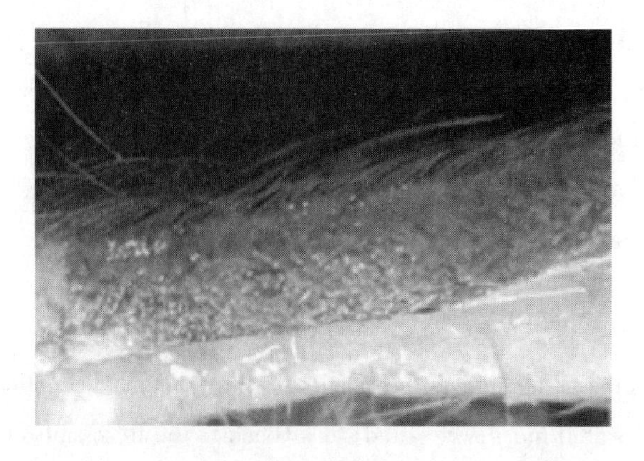

Figure 3.2 State of hide with satisfactory soaking

Several factors or parameters should be considered in the soaking process. The factors include the quality of the water and long float, temperature, pH of the water, auxiliaries, mechanical action, time, etc.

3.2.1.1 The quality of the water and long float, temperature, and pH of the water

The quality of the water is important and needs to be addressed. If the water is significantly loaded with bacteria, as may be the case with directly abstracted ground water, unlike processed town water, the likelihood of damage is increased. Soaking float is typically 200%-300% on the weight of salted pelt, depending on the raw material and its state of cleanliness. The float length tends to be consistent, even when multiple soaking steps are employed. The removal of salt is a diffusion and osmotic pressure-controlled process, so the concentration of salt in solution determins the rate of salt removal, and hence changes of float are necessary. In addition, the options for changing float have already been addressed. One of the advantages of using only fresh water is that the requirement to include a bactericide may be reduced.

Raising the temperature of the water aids in the dispersion of the globular proteins and accelerates the soaking process. The temperature of soaking float must be limited to <30℃ because the denaturation/shrinkage temperature (T_s) of the pelt is ~65℃ at around neutral pH. It has been suggested that the arrector pili muscle can still contract when the raw pelt is placed in cold water. Hence, it may be better to use water at 20-30℃ rather than at 10-20℃ in order to avoid raising the angle of the follicle and thereby risk coarsening the grain. Typically, soaking is conducted without pH change: due to the combination of pelt at physiological pH 8 and water at pH 6 due to absorbed carbon dioxide, the bath is typically around pH 7. However, some tanners add alkali to the bath, usually sodium carbonate, to raise the pH to about 10 to assist the wetting action.

3.2.1.2 Auxiliaries

Auxiliaries include akali, salt, surfactants (detergents), lignocellulosic enzymes, soaking enzymes, biocides, etc. Alkali tends to loosen the hair and epidermis. Usually, 1-3 parts of caustic soda, or of a milder alkali, such as soda ash, are used per 1,000 parts soaked liquor so that the pH of liquor is controlled at 9.5 or so. The surface fibers of the skin will rapidly absorb it and swell so much that they distort the surface of the skin, block up the inter-fibrillary spaces, and prevent the water from reaching the inside if too much alkali is used. Thus the loose grain may be shown. A salt solution of 3% concentration assists in dissolving the unwanted inter-fibrillary protein and thus speeds up soaking. Salt in the soak accomplishes two purposes: it aids in the removal of the globular proteins, and it aids in disinfecting the soak solution and thereby decreasing the possible case of bacterial degradation. If the soaking operation is conducted in a drum, the liquor ratio is such that the quantity of water is low; consequently, the salt washed out from the cured hide is generally sufficient to maintain salt concentration at a proper level for effective removal of some of the soluble protein. In the case of dry hides or prolonged soaking systems, the addition of salt will permit the use of higher temperatures with decreased probability of bacterial damage.

Hides or skins contain grease glands, fatty components, and sebaceous grease. Among pigskin, cattle hide, sheepskin, and goatskin, pigskin, and sheepskin belong to fatty skins and are the state of wet-salted in curing. However cattle hide and goatskin are less fatty skins and are the state of dried, salt-dried, or wet-salted. The fatty components prevent the water and the water-soluble materials into skins and affect the result of the soaking process. Therefore, anionic and non-ionic surfactants (detergents) were introduced into the system of soaking and accelerated rehydration by wetting the fiber structure. But removing grease is less effective since the temperature of soaking is typically low and therefore the grease is not mobilized. The melting point of triglyceride grease is 40-45℃ so that the actions of degreasing agents are greatly reduced at lower temperatures. Triglyceride grease is associated with the fiber structure and is present in varying amounts dependent on the species of animal: it is effectively removed by alkyl poly (ethylene oxide) type non-ionic detergents. Sebaceous grease is better removed by anionic detergents. Currently, available anionic detergents can bind to collagen and survive into post-tanning, when they can adversely affect dyeing and water resistance. This ionic interaction is important in post-tanning.

Wetting agents are recommended in concentrations of 2-5 parts per 1,000 parts water, particularly if the hides or skins are very greasy and, therefore, difficult to be wetted. Modem techniques use enzyme preparations. Soaking enzymes include proteases, lipases, amylases, lignocellulosic enzymes, etc, which not only break down protein, particularly nonstructural protein, to enhance the rehydration process but also contribute to the opening up of the fiber structure and assist soaking. They have specified proteolytic action on the inter-fiber proteins. Wet-salted hide may be so treated with 0.15%-1% of such preparation at pH 9-10 for 3-5 hours. They may assist in giving a smoother, flatter grain. Another system favoured for dry hides uses enzymes with an optimum pH 4-5 and often with a little non-ionic wetting agent.

During the soaking processing, the major risk is from bacterial activity which can be controlled by the use of biocides. Therefore the biocides are introduced into the system for the quality of hides or skins and the result of soaking. The choice of biocide is determined by the requirements of the raw material and the product to be made and compatibility with other agents present, e.g. enzymes. In deciding whether the use of a biocide in soaking is advisable, the prevailing conditions must be assessed which include pH, time, temperature, and the state of hides or skins.

The use of biotechnology is also proposed as alternative to the traditional soaking agents to increase the efficiency and reducing the environmental impact. Such technology is based on a fermentation process using beneficial microorganisms (probiotics) and natural raw ingredients. The broth resulting of the fermentation (metabolites) increases the soaking effect and, often, there is no need of bactericides.

3.2.1.3 Mechanical action and time

Mechanical action vigorously circulates the water around the skin and helps the loosening of cementing fibers by kneading or flexing the skin in the water. The action of drumming is more violent than paddling. The vigour of the kneading or pummelling action will be increased with the diameter of the drum, the size of the load, and the speed of rotation, and will decrease with the amount of water added. Normal drum speeds for soaking are 3-5 r.p.m. and drumming may be intermittent, e.g. for 5 minutes every hour. Drumming may reduce soaking time by 50%. However, violent mechanical action is not recommended in the earliest stages of processing when the hides or skins are in a relatively stiff condition. Therefore it is essential that the first stage of rehydrating is conducted in a static bath for dried hides or skins. When the hides or skins have softened,they can be conducted with any mechanical action.

Time is important in conventional batch soaking. Salted hides usually require 6 hours or more to remove enough salt to ensure that the pelt is completely rehydrated in the center of the cross-section and down the hierarchy of structure.

In the case of dried hides or skins, 24 - 48 hours or more is required which depends on the thickness of hides or skins, the degree of drying, etc.

3.2.2 Degreasing

All animal skins containa certain amount of grease, which is shown in Table 3.1. It can be seen from the table that the content of grease of the sheepskins and pigskins is higher than that of bovine, buffalo, yak hide, and goatskin. The sheepskins and pigskins are regarded as fatty skin. The fatty substances include triglycerides, waxy organic esters and fatty acids who are partly to be found in fat cells surrounded by membranes and in the sebaceous glands.

Table 3.1 The content of grease of several animal hide or skin

Sort of animal hide or skin	Content of grease/%	Sort of animal hide or skin	Content of grease/%
Bovine hide	<2	Goatskin	3-10
Buffalo hide	<2	Sheepskin	30
Yark hide	<2	Pigskin	15

Excessive amounts of natural fat in animal skins interfere with uniform penetration of water-base auxiliaries and the reaction between collagen and these auxiliaries in the process of leather manufacture and directly affect the quality of the finished leather. Therefore, in order to improve the penetration of auxiliaries and react with the hide or skin, the grease should be removed from the animal skins. The purpose of the degreasing process is to remove the grease distributed in the inner and surface of animal skins, to partly remove inter-fibrilary substance and to weaken the connection between hide or skin and dependent tissue such as epidermis, hair follicle, etc.

Conventional degreasing methods include mechanical degreasing and chemical degreasing. The former is to remove the natural fat by defleshing machines. Mechanical degreasing operation can not only remove the subcutis and the grease distributed on the surface of animal skins, but also damage the adipose gland and the fatty cells which is good for chemical degreasing. The latter is to remove the natural fat by emulsification, hydrolysis, saponification, or organic solvent methods. Emulsification methods have been carried out to degrease raw skins, bated skins, or pickled skins by drumming or paddling with surfactants. By the hydrolysis reaction for ester bond between triglycerides and lipase or alkali, triglycerides are turned into glycerol and fatty acid. Then in the solvent degreasing methods, the organic solvent such as kerosene, petroleum ether, dichloromethane, etc, is adopted to dissolve the grease in leather. Single mechanical degreasing or chemical degreasing operations can't completely remove the grease distributed in animal skin, combination degreasing methods between mechanical and chemical degreasing or more chemical degreasing methods are applied in the process of leather manufacture. In addition, one or two degreasing operations can't also completely remove the degrease of animal skin so that the degreasing process should be carried out in more processes such as the single degreasing process, liming process, bating process, pickling process, etc.

In order to achieve the purpose of the degreasing, it is very important for the degreasing effect by the judgement of the sort of animal skin and the controlling of some parameters such as the type and consumption of surfactant, the consumption of lipase or alkali, and the pH of the solution, the temperature of degreasing solution, the degree and time of mechanical action, etc.

3.2.3 Unhairing

Unhairing is the process of removing the hair from the pelt and its function is to remove hair or wool, the epidermis, residual interfibrillary components (particularly dermatan sulphate) and to open up the fiber structure. Unhairing and liming are often linked because the traditional processes of hair dissolving and alkaline hydrolysis combine the process steps in one. However, they strictly ought to be thought as separate processes and the steps are increasingly commonly conducted separately in modern processing. Unhairing is traditionally one of the dirtier aspects of leather processing, the odour is created when typically sulfide is employed and the polluting load is generated. The traditional method of dissolving the hair is hair burning, that is the sulphide unhairing methods and used in almost all the processes of leather manufacture. In order to understand the unhairing process, the components, structure of hair and conventional unhairing methods are introduced.

3.2.3.1 The components and structure of hair

The components of hair is keratin which is different from other main proteins in skin in its elemental composition and are shown in Table 3.2.

Table 3.2 Elemental composition of skin protein

Protein	Total nitrogen/%	Total sulfur/%
Collagen	17.8	0.2
Elastin	16.8	0.3
Keratin	16.5	3.9

Depending on the application, keratins vary in structure and composition. Keratins contain less glycine and proline than collagen and no hydroxylysine, but the contents of cystine and tyrosine are high. Cysteine can be converted into cystine by oxidation which is shown in Figure 3.3. The reaction takes place in the pre-keratinised zone during the growth of hair which is an equilibrium that can be exploited in the removal of hair from hides and skins. Keratin is usually encountered in the β-form, which is a stretched helix, in the way collagen's fundamental structure is helical. It forms fibrils created from protein chains linked via disulfide bonds, which are the target of many of the unhairing techniques.

The structure of hair is shown in Chapter 2. From the viewport of the cross-section of hair shaft, the hairs conclude cuticle, cortex, and medulla. The cuticle is the outer surface structure of the air

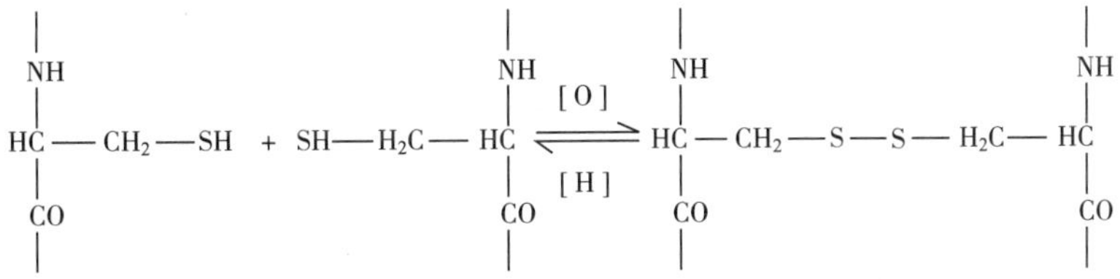

$$\text{HC(NH)(CO)}—CH_2—SH + SH—H_2C—\text{HC(NH)(CO)} \underset{[H]}{\overset{[O]}{\rightleftharpoons}} \text{HC(NH)(CO)}—CH_2—S—S—H_2C—\text{HC(NH)(CO)}$$

Figure 3. 3 Interconversion of cysteine and cystine

staple, is made of hard keratin containing a high concentration of disulfide links and is not fibrous in texture. The cuticle is composed of sheet - like cells that overlay each other, making it relativity chemically inert. The cortex is the inner structure of the hair whose structure is soft keratin containing less sulfur than cuticle and is fibrous. The fibrous structure isn't apparent because of its compact nature. The medulla is the central structure of the hair whose component is protein and not keratin since it does not have any sulfur. It is typically present in the cattle's hair. But the medulla may be missing in very fine hair or wool such as merino or merino cross wool, etc.

In addition, the hair also includes the bulb, the pre-keratinised zone, etc. The bulb contains the papilla where the hair grows. The structure of the bulb is protein, but not keratin. Young hair is still attached to the base of the follicle. Mature or club hairs are shed two or three times a year. The pre-keratinised zone is the region in the growing hair where the keratin is laid down and disulfide links are created. It is situated above the bulb and extends partway up the follicle.

3. 2. 3. 2 Conventional unhairing methods

Conventional unhairing methods are hair burning (hair-destruction) which refers to the practice of dissolving the hair with sulfide in the presence of lime. Hair burning occurs from the tip down to the surface and beyond, which usually occurs faster than penetration of the reagents through the pelt from the flesh side. However, once the reagents encounter the hair, they begin to dissolve the prekeratinised zone. The net result is that a plug of hair remains in the follicle which is commonly observed by microscope. Therefore pressure of handly or mechanically (operation "scudding") can be done in order to remove the residual hair from the follicle and clean the grain surface during the leather manufacture.

Hair-burning chemistry acts on the softer keratins of the cortex and epidermis. The presence of melanin in the cortex gives rise to the observed hair color, which is alleged to influence the rate of degradation. Results of some research show that black hairs dissolve more slowly than white hairs. However, the melanin is released into the solution and remains as a component of the "scud", residual pigmented solids, which can affect the grain quality by being driven into the surfaces of the pelt by mechanical action, causing uneven discoloration. The cuticle is typically not dissolved under the normal sulfide concentration and time conditions of unhairing. The scales flake off from the degraded cortex, to remain as "scud", suspended in solution. The medulla is not attacked by the unhairing chemicals, so it merely breaks off, weakened by the hydrolytic effect of high pH.

Sulfide unhairing is one of the most conventional hair burning methods, which is sub-classified as paint unhairing, hair burning unhairing in drum.

- Paint unhairing

Paint unhairing methods are often used for skins or hides with abundant fur. The washed/soaked skins or hides are piled in order to drain off surplus water, then are painted, swabbed, or sprayed on the flesh side with a "paint" which resembles thick "lime wash" and may be made from approximately 50 parts lime, 50 parts water and 5−20 parts sodium sulfide. A large quantity of paint may be applied in thicker or greasier areas of the skin. After painting, the skins may be folded along the backbone, flesh side in to keep the paint off the hair or wool, and piled or hung until the hair or wool is lose. The skins or hide may also be "paired", i. e. Stuck together in twos, flesh to flesh, and piled. The sodium sulfide and such lime as dissolves in water enter the skin from the flesh side, penetrate through the corium, and dissolve the young keratin cells, which enclose the hair roots. After piling, wool is pulled and graded usually by hand, while hair is scraped off with a cured, blunt unhairing knife on the beam or by an unhairing machine. In order to improve the efficiency of paint unhairing, paint operation also is replaced by grouting with a certain concentration solution of sodium sulfide and some auxiliaries. The hides or skins are placed on the automatic spraying line of the grouting equipment and the flesh side of the hides or skins is grouted with the solution of sodium sulfide and some auxiliaries, which is shown in Figure 3. 4. The latter operation is the same as that of the paint operation.

Figure 3. 4 The grouting equipment and relevant grouting operation

Depending on the thickness of the skins or hides, the tightness of the fiber structure, and the amount of flesh or fat left on the skin, the process with the above paint may take 5−12 hours or longer. If defleshing operation can be done before painting, the effect of dehairing operation is very good. Defleshing operation can remove some of the surface fat and flesh so that the paint can easily penetrate into the skins or hides.

- Hair-burning unhairing in drum

If the hair is little value, the hides or skins may be drummed intermittently in a relatively strong sodium sulfide, sodium sulfhydrate solution, such as 200% water on hide weight, 2%−5% sodium sulfide (whose content is 60%) and 2. 5% lime. The hair's fiber structure was destroyed or degraded within 15 minutes. After the reaction lasts one hour, the hair and epidermis are reduced to a pulp because they are primarily composed of keratin. The mechanism of a chemical reaction is the reducing nucleophilic sulfide ion attacks the disulfide bond of cystine so that the disulfide link is broken and creates a cysteine moiety and an anionic cystine disulfide moiety. The disulfide moiety is attacked by sulfide and is reduced to cysteine and the sulfide is converted into polysulfide, which is shown in Figure 3. 5.

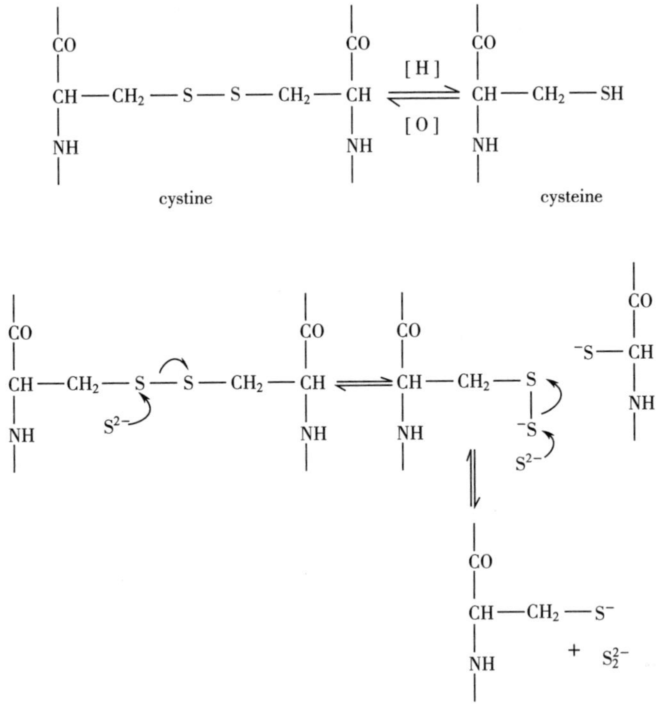

$$-CH(CO-)(NH-)-CH_2-S-S-CH_2-CH(CO-)(NH-) \underset{[O]}{\overset{[H]}{\rightleftharpoons}} -CH(CO-)(NH-)-CH_2-SH$$

cystine　　　　cysteine

Figure 3. 5　Mechanism of chemical reaction of hair burning

During the hair-burning unhairing, the hides begin swelling gradually, a certain amount of salt may be added and part of the sodium sulfide may also be replaced with sodium hydrosulfide in order to limit the swelling of skins or hides. Less swelling will give fine grain. When the hairs are completely dissolved, the drum is run intermittently for liming. The unhairing and liming are carried out in the same bath for 16-18 hours.

- Oxidative unhairing

Oxidative unhairing is a technology that has been used commercially based on sodium chlorite, which is conducted under acid conditions. The reaction involves the changing of an oxidation state to both higher and lower oxidation states.

$$4ClO_2^- + 2H^+ \rightleftharpoons 2ClO_2 + ClO_3^- + Cl^- + H_2O$$

$$4R{-}S{-}S{-}R + 10ClO_2 + 4H_2O \rightleftharpoons 8R{-}SO_3^- + 8H^+ + 5Cl_2$$

As shown in the reaction, the actual unhairing reagent is chlorine dioxide gas and then one of the products of the reaction is chlorine. Because of the strong oxidising chemicals involved, the process can not be conducted in wooden vessels, so stainless steel vessels have to be employed.

Other oxidizing agents that have been proposed are hydrogen, sodium, and calcium peroxides and peracids. For example, the options frequently appear in the literature, but none has obtained industrial acceptance.

3.2.4 Liming

Liming process is one of the most important processes during leather manufacture. The results of liming process directly affect the quality of the finished leather. An old saying is "*High quality leather originates from the tank with lime*".

3.2.4.1 The purpose and mechanism of liming

The purposes of liming are to remove non-collagenous components, dermatan sulfate of hides or skins, to split the fiber structure at the level of the fibril bundles, to swell the pelt, to hydrolyze peptide bonds, amide sidechains and fat, etc. For removing the non-structural protein, the albumins and globulins within the fiber structure of the hide or skin, the hydrolytic action of the liming process is so effective that it dominates the opening up processes and loosen or split the collagen fibrils at the level of the fibril bundles. Non-structural proteins are degraded and removed by the action of alkaline hydrolysis. Depending on the degree of the effect, the splitting can be characterized as coarse or fine which allows better penetration of reagents, make more effective reactions or obtain soft, strong leather. In addition, triglyceride fat, both cutaneous and sub-cutaneous, is likely to be present in the pelt at the beginning of liming, particularly if the pelts have not previously been fleshed. The fat is hydrolyzed by saponification reaction during liming, which is shown in Figure 3.6.

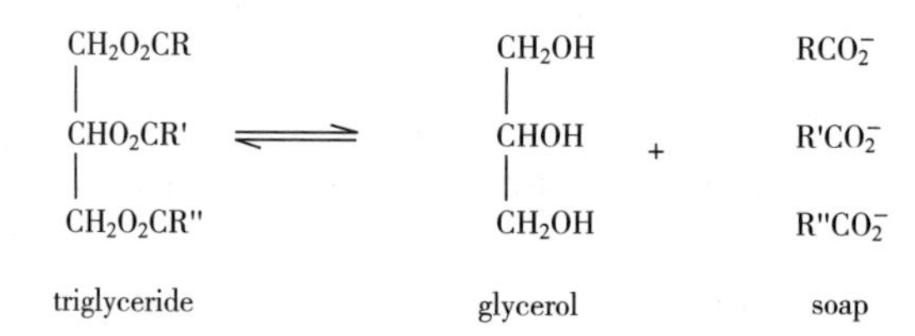

Figure 3.6 Hydrolysis of triglyceride

In the opening-up process, the fiber structure is damaged and chemical bonds, such as peptide bonds, amide sidechains, etc, are broken or hydrolyzed. Pelt can be swollen by charge effect, osmotic effects, lyotropy, etc. Charge effect is caused by breaking salt links and creating charges in the protein structure. Osmotic swelling is caused by the imbalance between the ionic concentration outside the pelt and inside the pelt. Lyotropic swelling is caused by disruption to the structure by species that can insert into the hydrogen bonding.

3.2.4.2 The liming operation

Liming process may be carried out either separately, or at the same time by immersing the hides and skins completely in the solution of calcium hydroxide (lime) and sodium sulfide. Unhairing commences with sulfide alone to avoid the possibility of immunising the kertain, but lime is soon added and so the combination of steps merely extends the period of treatment with two reagents. The hair is usually dissolved completely within a few hours, often 2–4 hours, then the liming is continued usually to a total of 18 hours. The pelt will be swollen in alkaline conditions, which can contribute to a problem if the rate of swelling is faster than the rate of hair degradation. If the pelt swells quickly, an additional portion of the hair shaft becomes hidden by the follicle in the swollen grain layer. Consequently, it is suggested that the liming process is placed after the hides or skins are dehaired and the minimum amount of water can be adopted in the hair burning.

Liming process is usually carried out with hydrated lime [$Ca(OH)_2$] and sodium sulphide (Na_2S) or sodium hydrosulphide ($NaHS$), which has a less pronounced plumping effect. The value of using lime in the process is that the pH is controlled only by the solubility of the sparingly soluble alkali.

Based on raw weight, the hides or skins may be immersed in pits in about 500%–600% times their weight of this liquor and lifted or agitated occasionally on in paddle or 200% in the slowly revolving drum, as in soaking for 12–60 hours, usually at 18–22℃. The relatively long float minimizes the possibility of abrasion damage to the grain. The minimum mechanical action is provided by the process vessel, either by processing in paddle driven vats or by running conventional drums slowly. Water will only dissolve a relatively small amount of lime, to give a clear solution (approximately 0.2 part lime per 100 parts of water). But the solubility of lime will be less if the temperature of the solution is raised. The solution causes the collagen fiber to swell by absorbing more water. These effects occur with all soluble alkalis. The stronger the alkalinity, the greater the effect.

In general, liming operation takes place between pH 12.0 and pH 13.0. When the hides or skins are introduced into the lime solution, the grain and flesh surfaces of the hides or skins are subject to solutions at pH 12.5 or so, the center of the hides or skins is still near neutral so that differential swelling occurs between the grain, flesh, and center of the skin in accord with the variation in pH existing at that time. The pelt can be swollen by three mechanisms, that is charge effects, osmotic swelling and lyotropic swelling. Charge effect are based on breaking salt links and creating charge in the protein structure. Osmotic swelling is caused by the imbalance between the ionic concentration outside the pelt and inside of the pelt. Lyotropic swelling is caused by disruption to the structure by species that can insert into the hydrogen bonding.

As liming proceeds, the difference of pH becomes less and the skins become uniformly swollen. The swelling curve is shown in Figure 3.7. During the lime process, lime or other alkali will disrupt some of the hydrogen bonding between the adjacent protein chains, which make available more acid and alkaline groups on the skin and can be proved by the change in the pH titration curve of the hide protein. Under the condition of alkaline, the hides or skins are opened up and the fiber bundles are splitted.

Peptide bonds, amide sidechains, and guanidino sidechains are hydrolyzed during the liming process. The longer the liming process goes on, the greater the degree of hydrolysis, and the greater the amount of damage to the protein components of the hide or skin, the more opening or splitting of fiber

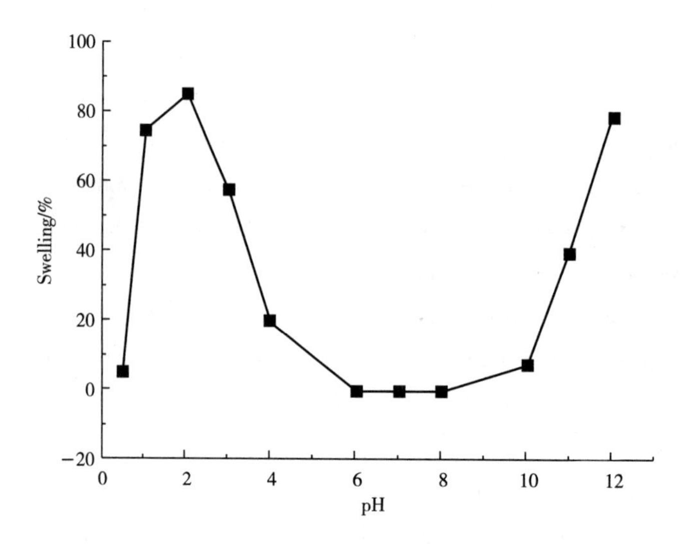

Figure 3.7 The swelling curve of hides or skins

bundles may take place in the hides or skins. Therefore, the prolonged swelling has a definite practical effect on the hides or skins with respect to its leather-making properties.

In the unhairing process, strong reducing agents, such as sodium sulfide can break down the sulfur-sulfur linkages of hairs so that the hair is destroyed progressively from the tip down toward the root. Sufficient alkali and strong reducing agents rapidly can destroy the hair after the beginning of the unhairing operation. But the hair roots may be still in corium after unhairing if the reaction is weak because of insufficient alkali or other causes. The residual hair roots may be continuously reduced by sulfides during the liming process.

3.2.5 Deliming

Thehide is a three-dimensional network of protein fibers which have absorbed alkali in the form of lime and other alkaline materials so that the pelt is always in the state of swelling during the liming process. Thus the collagen fibers can be loosened by removing the inter-fiber substances and other proteins dissolved in the alkaline solution which can give leather softness. But the process can also hydrolyze the collagen fibers and reduce the mechanical properties, such as tensile strength, tearing strength, and others. Therefore the pH value should be decreased after the completion of liming.

The process of deliming and bating are traditionally considered as inextricably intertwined because the former is thought of only preparing the pelt for the latter. Indeed, it is common for bating to be conducted in the deliming solution. However, from an analysis of the specific functions of the steps, it is useful to regard them as separate.

The functions of the deliming process can be lowering the pH and preparation for bating and mineral tanning, removing the lime and alkali from the liming process, and depleting the pelt. During the deliming process, the pH of the pelt is adjusted to pH 8.5 - 9.0 which is suitable for the biochemical reaction of protease. Pelt will be further adjusted to 2.5-3.0 in the subsequent pickling process. One of its main purposes of deliming process is to remove lime. It is part of the reaction of lowering the pH to neutralize residual alkali from the liming process. The change from pH 12. 5 to 8. 5 is accompanied by a significant degree of deswelling or depletion. The pelt appearance becomes less translucent, and more opaque as the water is lost and the fiber colour reappears. There is a net transfer of water from the inside of the pelt to the outside. This may seem a trivial restatement of the effect of the process step, that is the reversing of swelling, but it is of profound importance to opening up, because this is a rare occasion in the leather making process that such a net flow occurs.

Deliming can be conducted by washing and deliming agents. Washing can readily remove undissolved lime from the surface and some of the dissolved lime held between the fibers (Figure 3. 8). Continued washing with water will lower the pH by diluting the hydroxyl content and neutralizing the hydroxyl with the bicarbonate content in the water. Therefore the swollen hides or skins are washed. However, some of the lime or other alkali held chemically by the fibers have to be removed by acids or acid-producing salts such as weak acids, acidic salts, ammonium salts, etc. Weak acids are safer than strong acids because they do not cause the pH to drop low enough to cause acid swelling. The weak acid Lactic acid is one of the weak acid which produces a soluble calcium salt and has been used in industry. The application of aliquots of weak acids leads to the problem of too slow a rate of reaction, in the same way that it applies to strong acids. Acid salts might be used for deliming such as sodium bicarbonate, sodium metabisulfite, etc. Sodium bicarbonate has not been applied industrially because of cost and low solubility. The function of sodium metabisulfite is primarily to scavenge residual sulfide after sulfide unhairing. Ammonium salts are very soluble in water and are weak whose pK_a ($pK_a = pH + \log\{[NH_4]^+/[NH_3]\}$) is 9 or so and buffer around pH 9. It is common to use either ammonium sulfate or ammonium chloride in the leather industry.

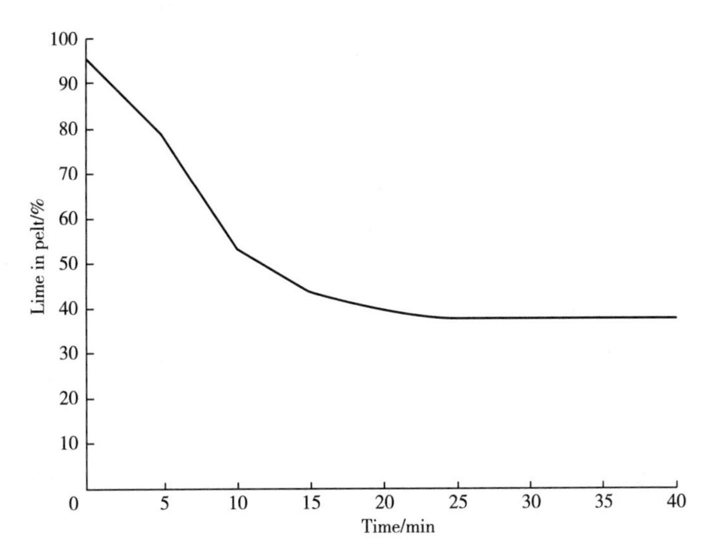

Figure 3. 8 Relationship between time of washing and elimination of lime in pelt

The hides or skins are usually drummed in 100%-200% water at 30-35℃ with 1. 5%-3% ammonium sulfate or ammonium chloride for 30 – 120min after washing. Calcium will be solubilized in ammonium sulfate or ammonium chloride so that the lime is gradually removed from the hide or skin by diffusion and the swelling effects on the hide during this operation are kept at a minimum. Penetration of the thicker parts of the hide or skin (neck, shoulder, or butt) is very much slower than in the bellies unless very vigorous mechanical action is given or the skins are very thin. During the deliming process, the tanner may cut the hide with a knife and use an indicator, usually phenolphthalein to determine the "lime steak" in the hide and thus measure the depth to which the deliming operation has proceeded. Whilst ammonia gas can be released into the working environment and ammoniacal nitrogen is discharged in the effluent during the deliming of the pelt using ammonium salts because ammonia is poisonous to fish and other aquatic life. Therefore there is an imperative to develop an alternative technology.

3.2.6 Bating

Bating is one of the critical processes during the process of leather manufacture and is a further step in the purification of the hide prior to tanning. The purpose of bating process is to remove the residual unhairing chemicals, to break down non-structural proteins, to weaken the action of elastic fiber, erector pill muscle, and muscle tissue on the tissue structure of hides or skins, to completely eliminate the swelling of pelt and to further loose collagen fibers. The residual unhairing chemicals include degradation products of pigments, hair root, hair root sheath, grease, epidermis, collagen, the "scud" on the surface of the skin, etc. Simultaneously the non-structural proteins can be degraded by general proteases because they do not have highly defined structure, even though they may be folded specifically. Thus bating process can give the leather fine grain, good softness, fullness, air permeability, and water vapour permeability.

Bating has been performed in different ways over the centuries. Traditional sources of enzymes have typically been animal faces, where the required component was the pancreatic or digestive enzymes. However, shortly after the beginning of the twentieth century, the first enzymes isolated from animal sources were made available for leather making by Otto Rohm. Now bating enzymes from bacterial fermentations have become available. Modern bating materials are based on sterile enzymes and are of two main types, that is pancreatic bates and bacterial bates. Pancreatic bates are made by using the digestive enzymes from the pancreatic glands of slaughtered animals. These are the glands which secrete enzymes onto the food in the stomach and eventually appear in the dung. These glands can be prepared in a sterile form and are then suitably mixed with fine sawdust and ammonium, sulphate or chloride. They appear as a coarse, yellowish, odour-free powder, easy to store and handle. In addition, bacterial bates can be obtained by bacteria. Appropriate bacteria are encouraged to grow in a suitable solution and soon become full of their digestive enzymes. Then the living bacteria are killed by sterilization, and the enzyme solution is prepared as a dry powder in a mixture of wood flour and ammonium salts.

In addition, the enzyme activity is affected by temperature, pH, time, origin of enzymes, and formulation. The reaction site of the enzyme is highly sensitive to temperature and the optimum temperature is usually 35–40℃. The protein may be denatured at a temperature slightly higher than the optimum temperature and the enzyme ceases to function. The effect of bating will be poor if the temperature is lower than the optimum temperature. The effect of pH is to change the charge on the protein and cause a change to the electrical field of the active site, with roughly similar changes on either side of the pH optimum. There is likely to be significant activity below and above the optimum pH, without the same catastrophic loss of activity due to damage to the reactive site on the enzyme. Time and the velocity of penetration of enzymes in the cross-section of hides or skins are also an important factor that affects the bating effect. Enzymes in the cross section of hides or skins hydrolyze the non-collagenous protein and the collagen. Commercial bate products are not pure enzymes and are not pure in terms of being a single enzyme. Therefore concentration and its formulation also affect the bating, offers of commercial bate products depend on bate formulation and the amount of water being used.

The pH of hides or skins is adjusted to 8.5–9.0 and washed before they are bated. The hides or skins are drummed in 100%–200% water at 37℃ with a 1%–2% addition of the powdered enzyme mixture (based on the enzyme activity and the needs of softness of the leather) and maintain the pH and temperature accurately of the float. The bating effect can only be monitored by traditional ways such as the thumbprint method for hides or skins, and the air permeability method for hides or skins. The effect of bating can remove the resiliency of the pelt when the interfibrillary proteins are degraded. When the pressure of the thumb finger squeezes the bated, opened-up fiber structure, the fiber structure doesn't spring back and leaves a thumbprint that recovers only slowly. The air permeability method for hides or skins relies on the effect of opening up on the ability of the pelt to allow the passage of air through it. The tanner folds the pelt to create a bubble by trapping air. Then holding the pelt tightly, the bubble is gradually made smaller, but the amount of air is maintained. This creates pressure within the bubble, forcing air from the inside to the outside, but only if the transmission is allowed by the degree to which the fiber structure is opened up.

3.3 Tanning operation

3-3

The tanning process is an important step in the process of leather manufacture which can convert the hides or skins into leather and improve the hydrothermal stability of hides or skins. The hydrothermal stability of leather depends on the tanning agents and its tannage. While tanning agents include mineral tanning agents (or in-organic tanning agents) and organic tanning agents.

3.3.1 Pickling and depickling

In general, the pH of the hides or skins and its float should be decreased before the mineral tanning agents are used for tanning process. The process of decreasing the pH of the hides or skins and its float is pickling. Therefore pickling refers to the treatment of the hide with salt and acid to bring the hides or skins to the desired pH for either preservation or tanning. Except for decreasing pH, the functions of pickling process are also to open up the collagen fibers, to further remove the non-fiber proteins and to partly dehydrate collagen, to protonate the carboxyl groups, and to be good for the preservation and shaving of the pelt, to complete the depleting of the pelt, etc. To further open up collagen fiber can increase the active group of collagen which is beneficial to the reaction between the tanning agent and collagen. To further remove the non-fiber proteins and to partly dehydrate of collagen can enlarge the room of collagen fibers which is helpful to penetration and reaction of tanning agents. To protonate the carboxyl groups can retard the reaction between the tanning agent and collagen, be suitable for the penetration of tanning agents and prevent over-tanning in the surface of hides or skins. The hides or skins have theoretically a purified network of hide protein at the end of the pickling operation.

$$^{+}H_3N—P—CHOO^{-}+H^{+} \rightleftharpoons {}^{+}NH_3—P—COOH$$

The pickling process is primarily conducted to adjust the collagen to the conditions required by the chrome or any other mineral tanning reaction. The traditional recipe for pickling includes 100% float, 6% salt, 0.6%-0.8% formic acid, and 1%-1.3% sulfuric acid based on the weight of limed pelt. The pickling is commonly carried out in the drums. The hides or skins are treated with 100% float. Then 6% salt is added to the drum and run to allow the salt to be soluble and absorbed into the pelt. The formic acid solution, and sulfuric acid solution (diluted 10 : 1 in water and cooled) are added to the drum from the hole of drum axle. In addition, the effect of pickling is affected by the conditions of pickling including the concentration of acid, the timing of the pickling step, the temperature of the float, mechanical action, etc. The offer of acid is determined by the pH required for the mineral tanning process. The pH of chrome tanning lies in the range of 2.5-3.0. However, the pH of other mineral tanning such as titanium, aluminum, zirconium, etc is lower than the one of chrome tanning. The timing of the pickling step depends on the thickness of the pelt, circumstances, and the requirement of different leathers. The effect of salt (sodium chloride) is to reverse or prevent acid swelling of the pelt. The salt concentration is one of the most critical factors in the quality of leather to be produced. The extent of the reversal depends on the salt concentration. The amount of salt depends on the float in the pickling and the hides or skins. It is customary to maintain the salt concentration at approximately 5% sodium chloride. The higher is the concentration of the solution of salt, the less swelling of pelts is observed. The temperature of the float has great influence/impact on the pickling process. Although higher temperature can speed up/accelerate the pickling process and be good at the penetration of acid in the pelt, the higher the temperature, the stronger the hydrolysis of hides or skin protein by acid so that the strength of finished leather is reduced and the possibility of loose grain is increased. In the production of leather manufacture, the temperature of pickling process should be kept 20℃ or so and lower than 25℃ by adding ice or the methods of connecting between running and stopping.

- Depickling

Occasionally, the tanner buying pickled hides or skins will want to adjust the pH to a higher value than that carried in the skins. To do this, the skins should be covered with brine and water containing the proper amount of salt. Then the pH can be raised. This procedure is seldom satisfactory because the pH of the solution may have little relation to the quantity of the acid remaining in the skins. The equilibrium conditions are seldom established and the usual results of such a procedure are inconsistent quality and such side effects as poor grain and hard leather. If a depickling system is to be used, it should be at least overnight to equilibrate the pH.

3.3.2 Tanning

The definition of tanning is the conversion of a putrescible organic material into a stable material that resists putrefaction by spoilage bacteria. Hides and skins are turned into leather by tanning. Tanning is done in the same large drums as used for bating and pickling, as soon as the later operation is complete. Chemical materials for tanning are tanning agents which cause some changes including appearance, handle, smell, shrinkage temperature/denaturation temperature, resistance to putrefaction by microorganisms, and other permanence, etc. Dried raw pelt is translucent and horny, but tanned leather dries opaque or may change in color, e. g. chrome tanning. By tanning, the leathers having some degree of softness are obtained.

Smoke, tanning materials obtained from the barks or leaves of trees, animal and fish oils, and certain salts are successively used as tanning agents. Certain salts are mineral tanning agents including basic aluminium sulfate, basic chromium sulfate, basic zirconium sulfate, etc. Aldehyde tanning agents such as glutaraldehyde, modified glutaraldehyde, tetrakis-(hydroxymethyl) phosphonium sulfate (THPS), oxazolidine, etc are used for tanning. In addition, a variety of synthetic chemicals are specially made for tanning and termed "syntan". Different tanning agents give leather different color, smell, and properties.

Hydrothermal stability is one of the most important indexes of tanning characteristics, which is the measurement of the resistance of a material to wet heat. In the case of collagenic materials, pelt, or leather, it is the effect of heat on water-saturated material. The thermal properties of collagenic materials are dependent on water content, hence when they are tested in this context they are required to be wetted to equilibrium arid therefore they are presented in a reproducible condition for testing.

The shrinkage temperature of pelt or leather is the most commonly quoted measurement of hydrothermal stability, certainly in technical publications and specifications, and is used as a sign of the hydrothermal stability of the leather. The shrinkage temperature has been assumed to be the breaking of hydrogen bonds in the triple helices, which would apply equally to raw and chemically modified (tanned) collagen, allowing the protein to undergo the transition from helix to random coil. Weir has demonstrated that, in the early stages of the transition between the intact helix and random coil structures of collagen, shrinking is a rate process and the rate is influenced by the moisture content: shrinking is independent of the soaking period once the collagen is hydrated to equilibrium, but drier specimens shrink more slowly. The rate therefore affects the value assigned to the shrinkage temperature: the slower the rate of shrinking, the higher the quoted shrinkage temperature. The kinetics of shrinking are affected by pH, the value drops at pH extremes and this is reflected in the thermodynamics of the reaction. The principle of the method is to suspend the test piece in water, in the form of a strip (the dimensions are immaterial since the property does not depend on sample size), then to heat the water at a rate of 2℃/min, according to the official method.

The shrinkage temperature of pelt drops from 65℃ to 40 - 50℃ after the pelt is limed and pickled. During the tanning process, some new cross-linking bonds are formed between tanning agents with adjacent peptide chains so that the stability of the structure of protein is strengthened and the shrinkage temperature of the leather is higher than the one of the pelt. The shrinkage temperature of the leather tanned by aldehyde, the leather tanned by vegetable tannin, the leather tanned by titanium sulfate, the leather tanned by zirconium sulfate can obtain 75-90℃, 74-85℃, 80-90℃ and 90℃ or so respectively. The leather tanned by basic chromium sulfate can resist to boiling water and has good comprehensive properties (Table 3. 3).

Table 3. 3 The appearance, shrinkage temperature and other properties of the leather tanned by different tanning agents

Pelts or leathers	Appearance	T_s/℃	Effect of water
Pickled pelt	White, opaque, soft	40-50	Reverts to raw
Leather tanned by vegetable tannin	White or brown, opaque, soft	74-85	Vegetable tannin isn't easy to be washed out
Leather tanned by chromium sulfate	Blue, opaque, soft	95-120	Cr (Ⅲ) isn't easy to be washed out
Leather tanned by aluminium sulfate	White, opaque, soft	70-74	Al (Ⅲ) washed out, reverts to raw
Leather tanned by titanium sulfate	White, opaque, soft	80-90	Titanium (Ⅲ) isn't easy to be washed out
Leather tanned by zirconium sulfate	White, opaque, soft	90 or so	Zirconium (Ⅳ) isn't easy to be washed out
Leather tanned by aldehyde	White, opaque, soft	75-90	Aldehyde isn't washed out

3. 3. 3 Chrome tanning

3-4

All leather was virtually made by "vegetable" tanning, i. e. using extracts of plant materials up to the end of the nineteenth century. Now the most common method of tanning is chrome tannage. Almost 90% of the world's output of leather is tanned in this way. The development of chrome tanning can be traced back to Knapp's treatise on tanning of 1858, in which he described the use of chrome alum: this is referred to as the "single bath" process because the steps of infusion and fixing of the chromium (Ⅲ) species are conducted as consecutive procedures in the same vessel. It is usually accepted that chrome tanning started commercially in 1884, with the new process patented by Schultz: this was the "two bath" process, in which chromic acid was the chemical infused through the hides or skins, conducted in one bath, the pelt was then removed to allow equilibration (but no fixation), then the chrome was simultaneously reduced and fixed in the second bath. However, chrome tanning may have a longer history, as proposed by Thomson, but the clearly observed change in tanning technology occurred at the beginning of the twentieth century. It was soon recognized in the industry that the new reaction was faster and more reliable than the then current vegetable tanning reactions, and so after several thousands of years using vegetable tannins, the global leather industry was converted to chrome tanning. The chrome tanning can be accomplished in a much shorter time (a few hours) than prior classical methods and produces a leather that combines to best advantage most of the chemical and physical properties sought after in the majority of leather uses.

The mechanism of the chrome tanning is the matching of the reactivity of the chromium (Ⅲ) salt with the ionized carboxyl groups of the collagen and form coordinate bond/cross-linking bond. The rate of reaction between chromium (Ⅲ) and unionized carboxyl groups is so slow that it can be neglected. The bond energy of the coordinate bond formed between chromium (Ⅲ) salt with the carboxyl groups draw a parallel with the one of covalent bond which strengthen the fiber network of the pickled pelt and convert the pelt into the chrome leather.

3.3.3.1 The chrome tanning agent and its chemical reaction

The main chemical composition of the chrome tanning agent is basic chromium sulfate or chromium (Ⅲ) salts, which are typically prepared from chromium (Ⅵ) compounds (commercially derived from chromite ore). Chromite ore is roasted in rotary kilns at 1200℃ and is converted into dichromate in the presence of alkali and oxygen. Reduction action converts dichromate into a baisc chromium (Ⅲ) under the condition of sulphuric acid and reducing agent such as glucose, sulfur dioxide (SO_2), etc. Reduction is illustrated in the following equation. It is crucial to ensure the reducing reaction is complete so that the leather is not contaminated by chromiun (Ⅵ).

$$H^+ + 2CrO_4^{2-} + 3SO_2 \rightleftharpoons 2Cr(OH)SO_4 + SO_4^{2-}$$

$$4K_2Cr_2O_7 + 12H_2SO_4 + C_6H_{12}O_6 \longrightarrow 8Cr(OH)SO_4 + 4K_2SO_4 + 6CO_2 + 14H_2O$$

Sugar-reduced basic chromium (Ⅲ) tanning salt is obtained by organic reductants. Other reducing agents were also used industrially, particularly sulfite, metabisulfite and thiosulfate. The difference between organic reduction and inorganic reduction is the presence of organic salts as residues of the breakdown of the organic molecules. These salts can act as complexing ligands, which change the composition of the ligand field around the chromium (Ⅲ) and hence lead to some effects on the leather.

Chromium (Ⅲ) salts are stable in the range pH 2-4. As the basicity of the solution of Chromium (Ⅲ) salts changes, the state of Chromium (Ⅲ) salts also changes. They will precipitate at higher pH values and the pH is the precipitation point of Chromium (Ⅲ) salts, which is shown as Figure 3.9.

$$Cr^{3+} \underset{}{\overset{OH^-}{\rightleftharpoons}} [Cr(OH)]^{2+} \overset{OH^-}{\rightleftharpoons} [Cr(OH)_2]^{+} \overset{OH^-}{\rightleftharpoons} Cr(OH)_3$$

Figure 3.9 The state of Chromium (Ⅲ) salts

"Basicity" is used to express the status of tanning salts: it is a way of expressing the degree to which a metal salt has been basified. The bigger is the basicity of the chromium salt, the higher is the molecular weight. The percentage basicity is defined by the following expression:

$$\text{percentage basicity} = \frac{(\text{total number of hydroxyl groups}) \times 100}{(\text{total number of metal atoms}) \times (\text{maximum number possible per metal ion})}$$

The basicity of chrome tanning agent can be calculated from the empirical formula $Cr(OH)_2^+$. In addition the chromium (Ⅲ) complex has some ligands such as H_2O, sulfate, etc. By basification, improving the temperature, enlarging the float ratio and extend the reaction time, etc, hydrolysis reaction occurs in the the chromium (Ⅲ) complex and release hydrogen ion, which is shown in Figure 3.10. Two or more chromium (Ⅲ) complex become the chromium dimer or the complex structure by olation and oxolation reaction so that hydroxy bridges, oxygen bridges are formed, the molecular structure of the chromium (Ⅲ) complex can be enlarged and the molecular weight increases, reactivity of the chrome species with the ionising groups carboxyl groups on the adjoining collagens and form the cross-linking bonds, which are shown in Figure 3.11 and Figure 3.12.

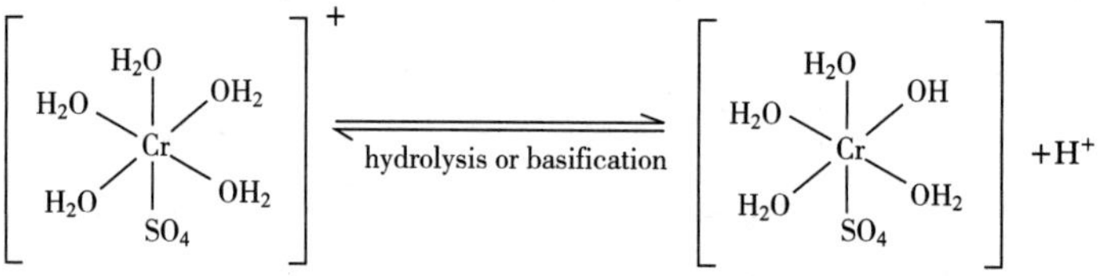

Figure 3.10 Hydrolysis reaction of the basic chromium (Ⅲ) complex

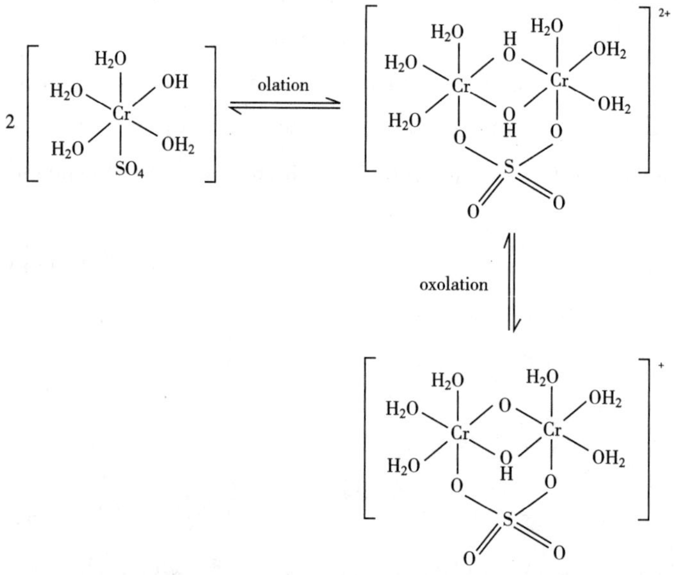

Figure 3.11 The olation and oxolation reaction of the basic chromium (Ⅲ) complex

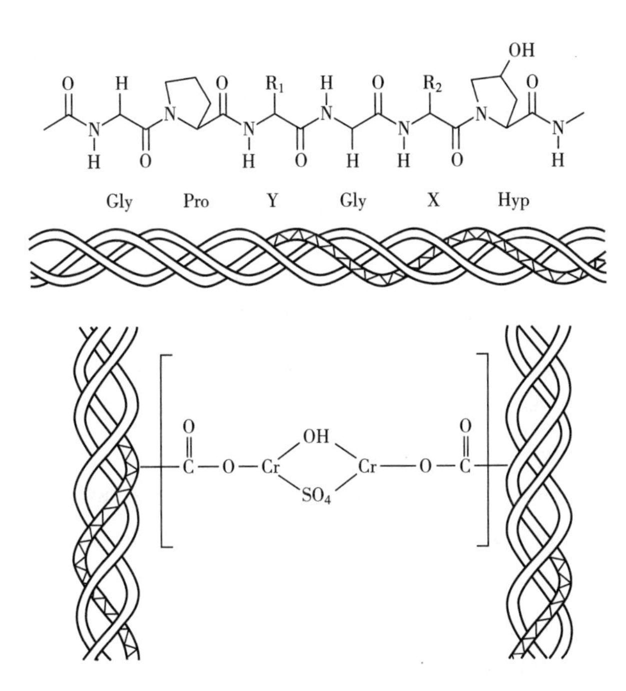

Figure 3.12 The cross-linking reaction between the basic chromium (Ⅲ) complex with the collagen fibers

3.3.3.2 Conventional chrome tanning operation

In chrome tannage, the pickled pelt is treated with basic chromium (Ⅲ) sulphate. Each chromium (Ⅲ) complex molecule reacts with several carboxyl groups of collagen to produce cross-linking between different molecule chains of the collagen, which strengthens the fiber network of the pickled pelt and converts the pelt into chrome leather.

The process of chrome tanning includes two stages. The first stage is the permeation process, that is the chromium tanning agent enters into the hide or skin and keeps even distribution in the section of the hide or skin at the start of tanning. The low pH value of the float, masking agents and the chromium salts with low basicity are important factors aiding to chrome penetration in the pelt by reducing the reactivity of the complexes. The masking agents include formate, acetate, oxalate, malonate, maleate, succinate, phthalate, adipate, etc, which reduce the charge on the chromium complex and reduce the initial affinity of the chrome complex for collagen carboxyls.

The second stage is binding reaction, which is coordination reactions occur between the carboxyl groups of collagen with chromium salts by basification, improving the temperature, enlarging the float ratio and extending the reaction time, etc.

Therefore, the process of chrome tanning is normally carried out in a pickling bath (or a part of the pickling bath). When the hides or skins have attained an equilibrium pickling about pH 3, the chrome tanning agent (33% basicity) is added. The tanning commences with the pelt surface at pH about 3.0 and a basicity about 33%. At this low pH, the hide or skin protein has little reactivity with chromium because its ionization is repressed. The chromium salt will enter the hide or skin and be fixed upon the grain and flesh layers. As the tanning proceeds, more tanning materials are deposited on the grain and flesh, and less will be deposited in the center, in accordance with the normal diffusion pattern.

After the chrome has penetrated the interior of the hide or skin (as shown by cutting the hide or skin and observing a uniform blue-green cross-section), the basicity of the liquor is increased by the controlled addition of alkali in order to fix the chrome. This is known as basification and must be carried out very carefully to avoid the dangers of precipitation of the unfixed chrome still in solution. Upon basification, the pH of the tannage is gradually raised, and the reactivity of the protein is greatly increased with increased pH, and more fixation of the tanning agent will take place. With the increase in basicity, two or more chromium atoms may be associated with one another through the common sharing of hydroxyl groups to form a larger complex, which is called olation. The higher the basicity is, the larger will be the complexity of the solution. The crosslinking reaction between adjoining chains of the protein will result.

After the pH of tanning bath has attained 3.8-4.2 for 30-60 minutes, a float of hot water (60-70℃) is added to the drum to raise the temperature to about 40℃. This will promote the fixation of the chrome complex by the hide protein, and reinforce the hydrolyzation and olation of chrome complex to a greater extent thereby producing more cross-linking between collagen molecules. The higher the temperature is, the more quickly the reactions take place. Care must be taken because excessively high temperature may cause uneven tannages, drawn grain, and green color.

As the tanning proceeds at the higher pH and temperature, the fixation and the crosslinking of chrome complexes to pelt continue. These reactions will result in a full tanning and a high shrinkage temperature. After running for another 4-6h, it is left overnight. Next morning it is run for 30min, and shrinkage temperature and pH value, as well as Cr_2O_3 content in effluent, are measured to check the degree of tanning. Then the leathers are horsed up.

In conventional chrome tannage, sodium bicarbonate is often used as a basification agent. Usually, it is dosed towards the end of tanning in several portions in a total amount of 0.8%-1.2% based on the weight of the pelt and achieve the final pH of 3.8-4.2 in the float. In addition Magnesia (Magnesium oxide), Calcium carbonate and dolomite can act as self-basifying salts. Dolomite is an equimolar mixed complex of calcium carbonate and magnesium carbonate, available naturally as dolomite rock and patented by the Bayer Company as a basifying agent for chrome tannage. The term self-basifying in this context means that the powdered dolomite is mixed with the 33% basic chrome tanning powder, so that at the end of the reaction the pH is a little below 4.0 and there is no need for the tanner to add base separately. Calcium carbonate decomposes uniformly in acid. The rate of reaction depends on the particle size: the faster the reaction, the higher the pH reached in the solution. In many cases, the solution pH is high enough to require the product to be added in aliquots-defeating the idea of self-basification. However, the high concentration of calcium ions in the solution may lead to the formation of calcium sulfate, by reaction with the sulfate present from the chrome tanning salt and (perhaps) from the pickling acid. This may result in the formation of sparingly soluble crystals within the grain surface, giving rise to an effect analogous to limeblast.

Magnesium oxide acts as a self-basifying agent and may also be used. Due to its low solubility, this agent reacts very slowly with acids. It can be added at the start of tanning in an amount of 0.6%–0.8% on pelt weight, gradually increasing the float pH up to 3.8–4.2. In a conventional procedure, the tanning and basification operation should be finished at a float temperature of 35–40℃.

3.3.4 Other mineral tanning

Except for the chromium (Ⅲ) tanning, other mineral tanning agents are used for the tanning process, which include aluminium salts, titanium salts, zirconium salts, iron salts, etc. They can give leather a certain hydrothermal stability.

3.3.4.1 Aluminium tanning

Aluminium (Ⅲ) tannage has been used in leather making for thousands of years. Aluminium (Ⅲ) salt is an acidic complex with a greater tendency to hydrolyze to basic salts than chromium (Ⅲ), especially at low concentrations. It also has a lesser tendency to olate and thereby to form discrete basic salts: it readily polymerizes to an Al_{13} species and precipitates. Sulfate does not complex with aluminium (Ⅲ) in the dry salt, as it does in basic chromium (Ⅲ) sulfate; the interaction is much more electrostatic. Aluminium (Ⅲ) is stable at pH 4, where the basicity is 30% – 35%, helped by the presence of a masking agent, otherwise, it is likely to precipitate at pH<4. Masking effects can be summarised by the following series: glucoheptonate>gluconate>oxalate>citrate>malonate>lactate>tartrate>succinate>acetate>glycolate>formate>sulfate.

Masking may be applied to aluminium (Ⅲ) sulfate, to allow the creation of a more reactive salt at pH approaching 4. Alternatively, basic chloride salts are available. They can be more basic because chloride ions are one of the few ligands that can create true covalent complexes. For this reason, the chloro-complexes are typically less reactive than those based on the sulfate, and additional sulfate in the tannage is beneficial.

Al (Ⅲ) interacts with collagen carboxyls, but in a reaction that is much more electrostatic than covalent so that the leather retains a significantly cationic character: in this state, the fiber structure is vulnerable to collapsing and sticking together, producing thin, hard leather or, in the case of pickled pelt, a tendency to form irreversible adhesions.

The weak interaction between aluminium (Ⅲ) and collagen means that the shrinkage temperature from a tanning process is barely raised. By the use of masking, the aluminium (Ⅲ) tanning process can be conducted at pH 4, yielding leather with a shrinkage temperature of about 80℃. Despite the obvious inadequacies of aluminium (Ⅲ) as a tanning agent, there are some modern uses of aluminium (Ⅲ) salts.

3.3.4.2 Titanium tanning

Titanium (Ⅳ) salts have a similar affinity for collagen as aluminium (Ⅲ), in part due to some similarities in properties, such as the acidity of the ion and the tendency to hydrolyze and precipitate as the pH is raised above 3. The interaction with collagen carboxyls is similarly electrovalent, rather than covalent. However, one difference is the greater filling effect of Ti (Ⅳ) salts due to the polymeric nature of the salts, which produces softer leather. The empirical formula of the cation, the titanyl ion, is TiO^{2+}: in reality, the ion is a chain $(—Ti—O—Ti—O—)_n$ with an octahedral ligand field. The weak chemical interaction with collagen results in shrinkage temperatures of 75–80℃. The resultant leather is initially colorless, although it tends to go pale yellow on aging, owing to the creation of charge transfer bonding; this is a change from hydrogen bonding to electrostatic bonding.

Titanium (Ⅳ) salts were widely used for the tanning of all kinds of leather such as hat banding leather, sole leather, etc. In addition, titanium (Ⅳ) salts were used in the semi–titan: starting with vegetable–tanned leather, it was retanned with potassium titanyl oxalate, to give a very hydrothermally stable leather, which is also capable of resisting the effects of perspiration because the elements of the combination tannage are rendered non–labile.

3.3.4.3 Zirconium tanning

Zirconium (Ⅳ) salts are expected to perform in a similar way to Ti (Ⅳ) because of their position in the periodic table. They have been available commercially since the middle of the twentieth century: hock has comprehensively reviewed the chemistry and tanning properties. The tanning reaction is similar to Ti (Ⅳ), yielding slightly higher shrinkage temperatures, but the color is stable and makes white leather with a full handle.

The chemistry of the complexes is more like chromium (Ⅲ) than titanium (Ⅳ). It can be seen that the 50% basic ion is a tetrameric species, where the metal is eight–coordinated, with u–hydroxy bridges between metal ions.

Like titanium (Ⅳ), the unmasked zirconyl ion is acidic and unstable to hydrolysis. The conventional way of applying it is to use as little water as possible for the penetration stage, then add water to cause polymerization by hydrolysis, in the same way as chromium (Ⅲ).

The increase in molecular weight increases the astringency of the species, driving it into the substrate, which results in fixation. It has been claimed, by deactivating the carboxyl groups and amino groups of collagen, that zirconium (Ⅳ) complexes do not interact with the carboxyls and that they interact significantly with the amino groups. However, the mechanism is unclear, since zirconium (Ⅳ) can coexist as cationic, neutral, and anionic species, by interaction with counterions, particularly sulfate and masking agents such as citrate. Significantly, in reacting with collagen, counterions are displaced from the zirconium complexes.

Zirconium (Ⅳ) tanning has been referred to as the mineral equivalent of vegetable tanning because of the presence of so many hydrogen bonding sites on the complex. This is a significant contributor to the binding mechanism in tanning. The large size of the complexes produces a filling tannage (a characteristic of vegetable tanning) and the commonest use for this chemistry is as a retannage, particularly to stiffen (tighten) the grain. Like vegetable tanning, zirconium (Ⅳ) tanned leather is hydrophilic because of the hydrophilic nature of the tanning species, as shown in Figure 3.13.

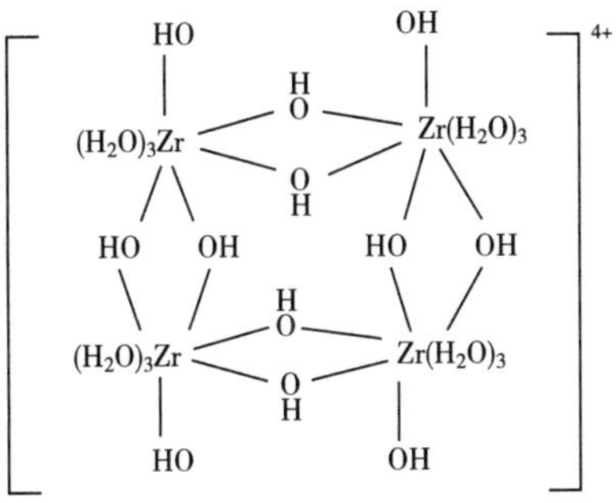

Figure 3.13 Structure of 50% basic zirconium (Ⅳ) ion

3.3.4.4 Iron tanning

Iron tannage took place in Germany in the mid part of the twentieth century when the availability of chromium was restricted. In 1921, Mr Hou Debang's PhD "Iron Tannage" was specially serialized by The Journal of American Leather Chemists Association and published in full.

Tanning with iron (Ⅱ) salts is preferred to iron (Ⅲ), primarily due to the oxidizing power of Fe (Ⅲ). The applications are limited because the lack of true complexation with collagen carboxyls means that the resulting shrinkage temperature is much the same as aluminium (Ⅲ). A disadvantage for workers is that they suffer from a metallic taste in the mouth, due to absorption of the metal salt through the skin when they handle wet leather.

Iron (Ⅱ) sulfate complexed with ethylenediamine and masked with dicarboxylate can produce a shrinkage temperature of 76℃, but tartrate masking alone can give a shrinkage temperature of 85℃. In addition, iron salts, especially Fe (Ⅲ), have a high affinity for plant polyphenols, although this may be a disadvantage in tanning because the resulting color is typically intensely blue-black. Hence, these reactions have been exploited for thousands of years for making inks. The effect is often observed in factories using vegetable-tanned leather, when the leather becomes affected by black spots from the action of sparks from, for example, the continuous sharpening of splitting band knives. The treatment or "clearing" action is to use oxalic acid or EDTA (ethylenediamine tetraacetic acid disodium salt) to complex the iron and solubilize it.

3.3.5 Vegetable tanning

Long long ago, the leaves, nuts, barks, and wood of plants were used for leather-making because prehistoric man perhaps observed changes in hide or skin after it had lain in a puddle with plant material. Therefore vegetable tannange is one of the oldest tanning technologies, which can stabilize collagen in hide or skin against putrefaction. The main content of vegetable tannin is plant polyphenols which can react with protein, that is "astringency". Tannins are extracted from plant materials by water or organic solvents. The tanning function depends on the phenolic hydroxyl content and the molecular weight of vegetable tannins. Phenolic hydroxyls are capable of reacting with collagen and forming hydrogen bonding between the basic groups on sidechains, partially charged peptides and tanning agents. In addition, the molecular weight of vegetable tannins is an important factor for tanning

3-5

properties. Lower molecular weight compounds lack astringency, a higher molecular weight causes surface reaction and impedes penetration into the pelt. The ideal molecular weight of vegetable tannins is in the range of 500 and 3,000.

Based on plant polyphenols' structural characteristics, they can be classified into the following two groups: one is hydrolyzable tannins, with subgroups gallo-tannins and ellagitannins, the other is condensed tannins.

3.3.5.1 Hydrolysable tannins

The hydrolysable tannins are saccharide-based compounds, in which the aliphatic hydroxyls are esterified by carboxylate species carrying the pyrogallol group (1,2,3-trihydroxybenzene). The tannins are typically based on glucose, but there are variations in which derivatives of glucose can be the central moiety, such as sucrose. They can be separated into two groups: the gallotannins and the ellagitannins. In each case, the properties are dominated by the ease of hydrolysis of the ester linkages and the reactivity of the phenolic hydroxyl groups. The commonly used hydrolytic vegetable tannins are valonia, sumach, chestnut, tara, etc. The structure scheme of TARA tannin, and chestnut tannin is respectively shown in Figure 3.14 and Figure 3.15.

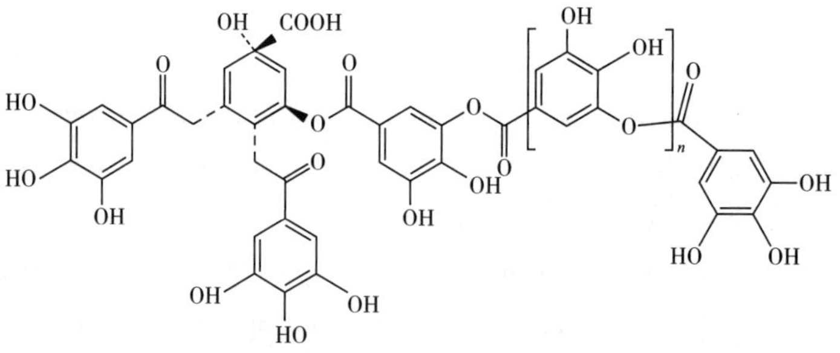

Figure 3.14 The structure scheme of TARA tannin (n=0,1,2)

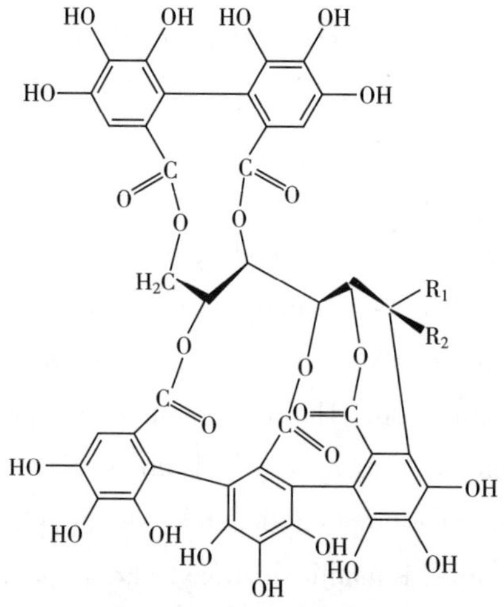

(R_1=H, R_2=OH or R_1=OH, R_2=H)

Figure 3.15 Major component of chestnut tannin

Gallotannins are the smaller group within the hydrolyzable tannins, which has less common polyphenols and the glucose core is esterified only with gallic acid. In addition, the bound gallate groups can be esterified themselves, at their phenolic hydroxyls: this is called depside esterification and illustrated in Figure 3. 16.

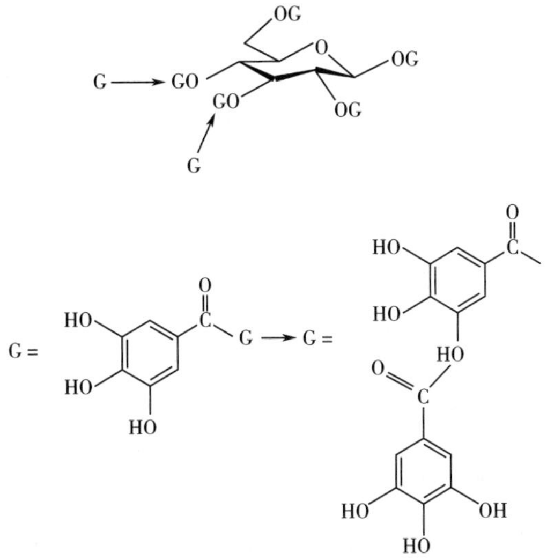

Figure 3. 16 Structures of the gallotannins, hydrolysable plant polyphenols

Variations in structures come from the degree of esterification of the glucose center and the degree of depside esterification. The astringency of any polyphenoldepends on the effective concentration of reaction sites within the molecule.

In ellagitannins compounds, the esterifying moieties include gallic acid, ellagic acid and chebulic acid (Figure 3. 17).

(a) ellagic acid

(b) chebulic acid

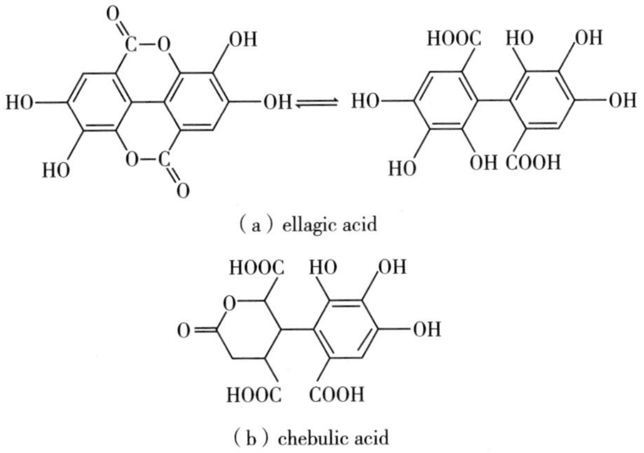

Figure 3. 17 Structure of ellagic and chebulic acid

Because the hydrolyzable tannins have a large number of phenolic hydroxyl groups and they are close together, they are highly astringent. However, the gallotannins are more astringent than the ellagitannins. The shrinkage temperature of the leather tanned by the hydrolyzable tannins can obtain 75–80℃. In addition, the presence of carboxylic acid groups can contribute to the reactivity of the hydrolysable tannins.

3.3.5.2 Condensed tannins

The condensed tannins have the flavonoid ring structure, as shown in Figure 3.18; the more common structures have a catechol group (3,4-dihy-droxy benzene) as the B-ring. Rings A and B are aromatic and ring C is alicyclic. The differences between tannins lie in the pattern of hydroxyls: position 3 in the C-ring is always occupied, in the A-ring position 7 is always occupied, but position 5 may or may not be occupied, in the B-ring position 3' and 4' are always occupied, but position 5' may or may not be occupied, as shown in Figure 3.19.

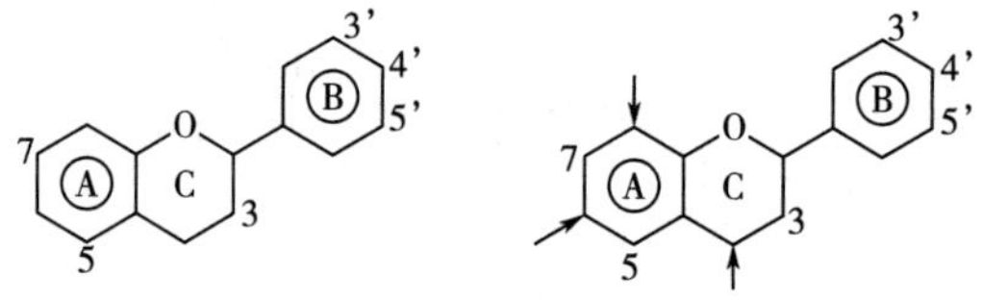

(a) hydroxylation positions　　(b) polymerisation positions

Figure 3.18 Flavonoid ring system, showing the positions for hydroxyls and the positions for polymerisation

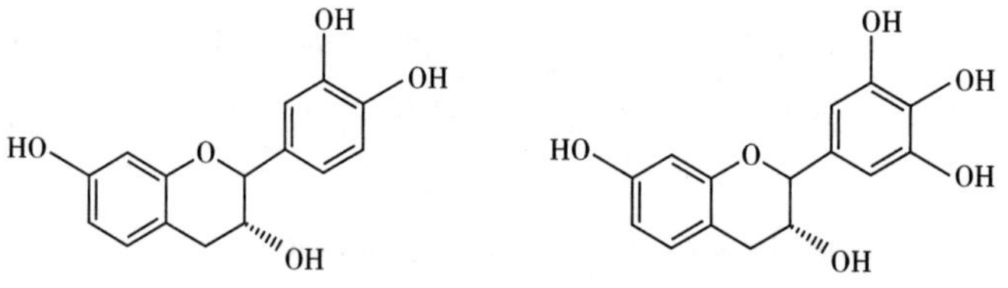

(a) profisetinidin　　(b) prorobinetinidin

Figure 3.19 Variations in hydroxyl patterns in the flavonoid polyphenols

The hydroxyl in the 3-position in the central C-ring may be cis or trans with the B-ring, as exemplified by gallocatechin, as shown in Figure 3.20. The alicyclic C-ring is the linking point for polymeriz ation and there is a variation in conformation, as shown in Figure 3.21.

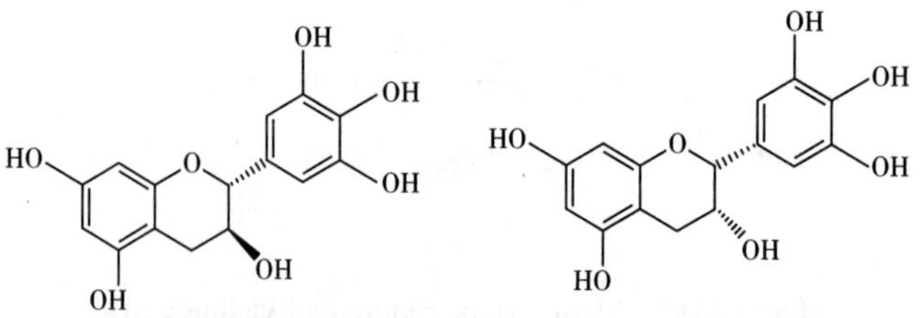

(a) (+)-gallocatechin　　(b) (–)-epigallocatechin

Figure 3.20 Stereo-chemistry in the flavonoid ring system

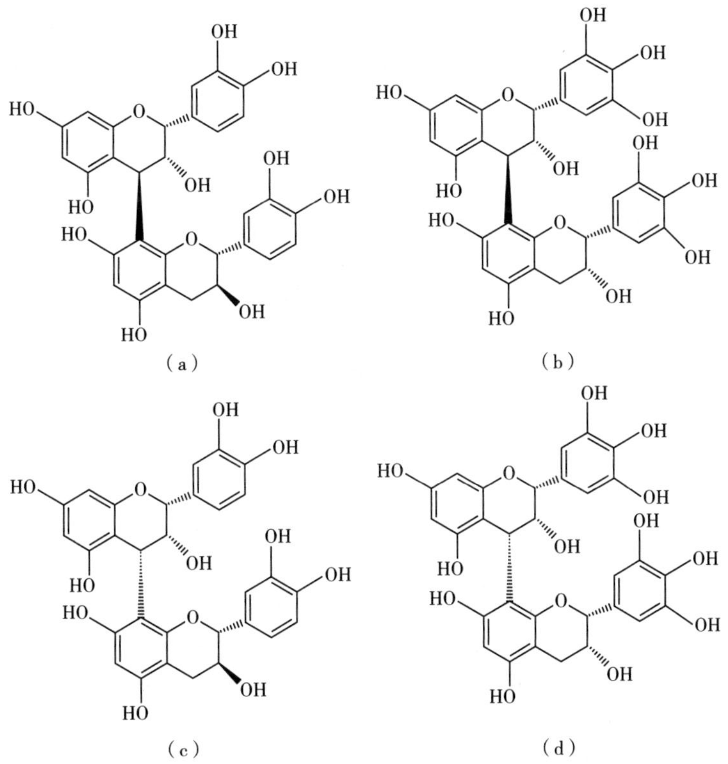

(a)

(b)

(c)

(d)

Figure 3.21 Conformations of procyanidins

The condensed tannins react with collagen by hydrogen bonds. In addition, they can react at the 5- and 7-positions via quinoid species, resulting in a covalent reaction at the collagen lysine amino groups. Because of this contribution to the fixation reaction, 5%-10% of the tannin is not removable by hydrogen bond breakers or organic solvent, so the pelt does not revert to raw after detanning. The shrinkage temperature of leather tanned with condensed tannin is 80-85℃. The proximity of the aromatic nuclei in the flavonoid structure means that free radical oxidative bond rearrangements can take place easily. Therefore, these tannins redden, creating a rapid color change on the leather surface: mimosa tanned leather exhibits this effect, darkening visibly after only minutes of exposure to sunlight in the air (Figure 3.22).

3.3.5.3 Vegetable tanning technologies

Historically, vegetable tanning was conducted in pits in a process referred to as "layering". A layer of the appropriate plant material, e.g. oak bark, was placed in the bottom of the pit, followed by a layer of hides or skins. Another layer of plant material was placed on the hides, followed by another layer of plant material; the alternate layering continued until the pit was full. Then the pit was filled with water.

The water leached the polyphenols out of the plant material and the dilute tannin diffused into the hides, converting them into leather. The dilute nature of the solution limits the reactivity of the tannins, allowing it to penetrate through the pelt cross-section. The reaction is slow, in part due to the static nature of the process, so that the requirement for hides to stay in the pit for a year.

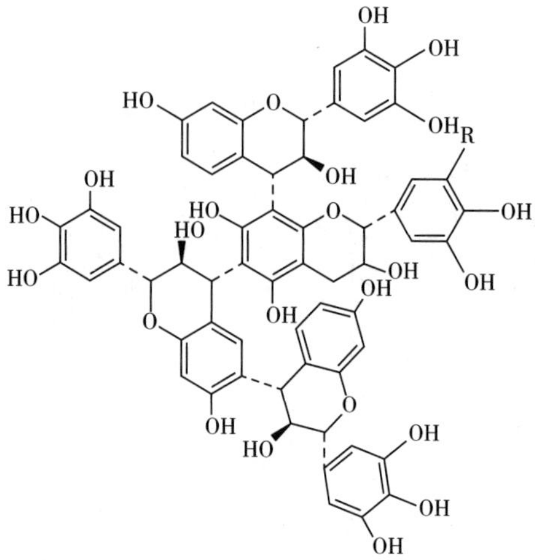

Figure 3.22 Structure of mimosa

As the development of both tanning agents employed and the equipment used for the tanning process, the vegetable tanning technologies were changed. Concentrating vegetable tannins have been obtained which can accelerate the tanning process and reduce the reaction times. But the astringency of vegetable tannins and its high concentration may overtan the surface of pelts, which forms a barrier to tannin penetration, and the tanning reaction within the cross-section stops. Therefore one approach to mitigate the effects of tannin astringency is to reduce the reaction rate and the easiest way to control the rate is to control the concentration of the tannin: the lower the concentration, the slower the rate of reaction. This is another example of the requirement from the tanner of balancing the rate of penetration and the rate of fixation. Thus four or more pits are used for vegetable tanning. The hides start in the first pit where the tannin concentration is low so that the reaction is slow and the tannin can penetrate through the cross-section. After a few days, the hides are ready to go into a more concentrated solution: faster reaction is allowed, because tannin fixation from the first bath has reduced the reactivity of the substrate somewhat. Some of the depleted solution is discharged to waste and the concentration is made back up with some of the solution from the second pit, ready to receive the second pack of rawstock. The process is repeated: the hides are moved to a third pit, and the first and second pits are 'mended' with higher concentration tannin solutions, the second pack is moved to the second pit and a third pack is introduced into the first pit. This process is continued until the leather is fully tanned and taken from the pits.

In order to facilitate the rapid penetration of vegetable tannins and their fixation with collagen, vegetable tanning can be conducted in pits or drums. In addition, higher offers of pretanning agents are used. Pretanning agents include non-swelling acids, syntans, or other hydrogen-bonding auxiliaries such as polyphosphates.

3.3.6 Other tannages

Besides mineral tanning and vegetable tanning, oil tanning, sulfonyl chloride tanning, syntan tanning, aldehyde tanning, etc are used for the tanning of pelt.

3-6

3.3.6.1 Oil tanning

Oil tanning is one of the oldest tanning methods and is used for making chamois or sheepskin leather. The tanning agents of oil tanning agents are unsaturated oils such as cod liver oil, whale oil, etc that involve filling and tanning properties by oxidation so that the shrinkage temperature of the leather is not raised significantly above the value of raw pelt. The reactions in the oil tanning process are not completely clear. One of view generally accepted is that the reaction is based on the formation of aldehydic compounds by oxidation, which has been used as an element of quality control.

The oil tanned leather is best known for its properties of holding water and can be useful for cleaning and drying washed surfaces. In addition, the oil tanned leather can resist microbial attack.

Take the oil tanning of sheepskin as an example, act pickled sheepskin pelt as materials, it was washed with fresh water in order to extend the opening up or loosening of the fiber structure and assist the penetration of the oil. Then the grain of the pelt was removed by splitting and the oil can penetrate into the pelt wet with water. The split pelt was repickled by 10% salt solution, the pH of the pickled float was adjusted to about 5.0 (the iso-electric point) and was pretanned by aldehydic agents such as glutaraldehyde, phosphonium salts, etc. Then the water content of the pelts was minimized by squeezing in order to allow oil penetration. Cod liver oil was added and was effectively hammered into the pelt by the action of a bank of large wooden mallets. The oil was oxidized by hot air at 40 - 50℃, then degreased with sodium carbonate solution or solvent and drying. The flesh surface of the oiltanned leather was buffed and conditioned under the condition of 20℃ and 65% RH.

3.3.6.2 Sulfonyl chloride tanning

Sulfonyl chloridehas been used as a sort of synthetic oil tanning, an alternative chamois process. They can react with the amino groups of collagen fibers. The leather is less colored and the shrinkage temperature rises to 80℃, giving similar properties to oil tanning.

3.3.6.3 Syntan tanning

The syntan refers to the range of synthetic tanning agents which is synthesized, patented and used in leather making during the Second World War. There are two steps in the synthesis of syntans: sulfonation and polymerization. These steps can be conducted in either order. The reaction of phenol or naphthalene and formaldehyde, concentrated sulfuric acid is shown in Figure 3.23 and Figure 3.24. Sulfonation can increase the solubility of the syntans in water and reduce the astringency. Polymerization can enlarge the molecular weight of the syntans and improve the tanning, and filling properties. Therefore the degree of polymerization and the number of sulfonic groups are connected with the tanning properties of the syntans.

In addition, the properties of the syntan depend on the nature of the monomeric precursors such as phenol, 2-naphthol, 2-methylphenol, 3-methylphenol, resorcinol, 4,4'-dihydroxydlphenyl sulfone, etc. The monomers can be polymerized by the crosslinkers such as formaldehyde, sulfonyl chloride, acetone, ethylene oxide, etc. The size or molecular weight of syntans relate to the ratio of crosslinkers to the monomers. The reactivity of the syntans depends primarily on the availability of phenolic

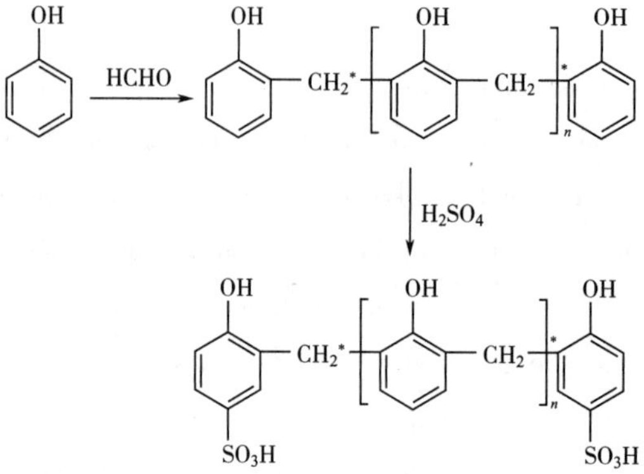

Figure 3.23 Polymerisation and sulfonation phenol

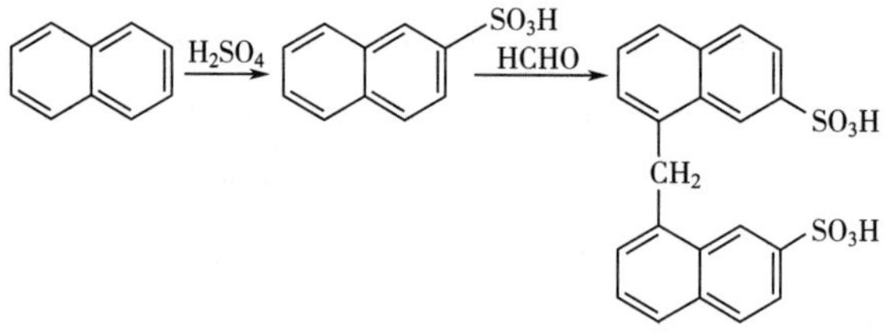

Figure 3.24 Polymerisation and sulfonation naphthalene

hydroxyl groups for hydrogen bonding. Another reaction is fixation by the sulfonate groups.

Based on the ratio ofphenolic hydroxyl groups and the sulfonate groups, the molecular weight of the syntan, the syntan is classified into auxiliary syntan, replacement syntan, and other syntan.

In general, auxiliary syntans have low molecular weight and few phenolic hydroxyl groups. According to the functions of the auxiliary syntans, they include dispersing syntans, bleaching syntans, neutralizing syntans, non-swelling acids, etc. Dispersing syntans can aid to solubilize or disperse syntans with higher molecular weight, vegetable tannins, dyestuffs and promote the penetration. Bleaching syntans can reduce the color of the surface of the leather. Depending on the reqirement of the charge on the leather during the retanning, dyeing and fat-liquoring process, neutralizing syntans can be used for neutralizing the chrome tanned leather or non-chrome tanned leather and raising the pH of the leather. Some auxiliary syntans can be used for non-swelling acids in the pickling process in order to reduce the consumption of salts and protect the environment.

Compared with auxiliary syntans, the replacement syntans have high molecular weight, a few phenolic hydroxyl groups, few sulfonate groups and more rigid rings which are synthesized by 4,4'-dihydroxydlphenyl sulfone, bisphenol A, naphthalene sulfonic acid, phenylsulfonic acid, formaldehyde, etc. They can be used for tanning pelts as solo tanning agents and replace vegetable tannins. The structure of a replacement syntan is shown in Figure 3.25. The shrinkage temperature of the leather tanned by the replacement syntans can obtain 80-85℃ whose reactive mechanism is similar to that of vegetable tannins. In addition, the replacement syntans have stronger filling properties, giving the leather good fullness and handle. Meanwhile, the replacement syntans can minimize the color of the leather, and improve the lightfastness of the leather.

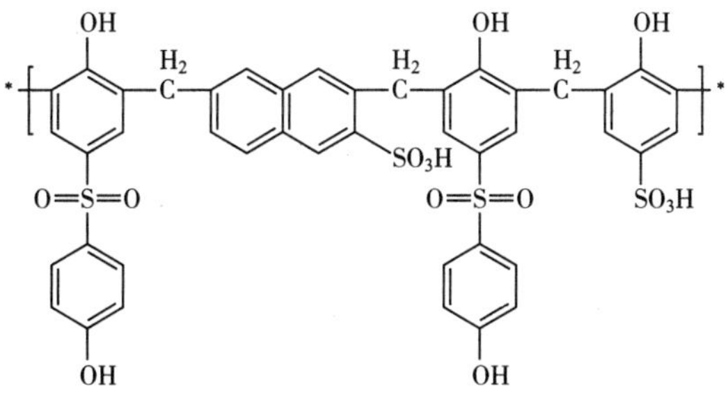

Figure 3. 25 The structure of a replacement syntan

During the retanning process, syntans with high or medium molecular weight can be also used as retanning agents and provide good filling effect and fullness.

Otherwise, according to theelectric charge of the syntans, they include anionic and amphoteric syntans. The amphoteric syntans contain acid and basic groups so that they are less reliant on pH variation for fixation.

3. 3. 6. 4 Aldehyde tanning

Aldehyde tanning is also one of the oldest tanning methods. For the preservation of the hides or skins and meats, smoking was used to treat the hides or skins by ancestors. Aldehyde is one of the main components of smoking. Aldehyde tanning agents include formaldehyde, glutaraldehyde, oxazolidines, phosphonium salts, and other aliphatic aldehydes which can react with the amino groups of collagen fibers and form covalent bonds. The mechanism of aldehyde tanning is shown in Figure 3. 26.

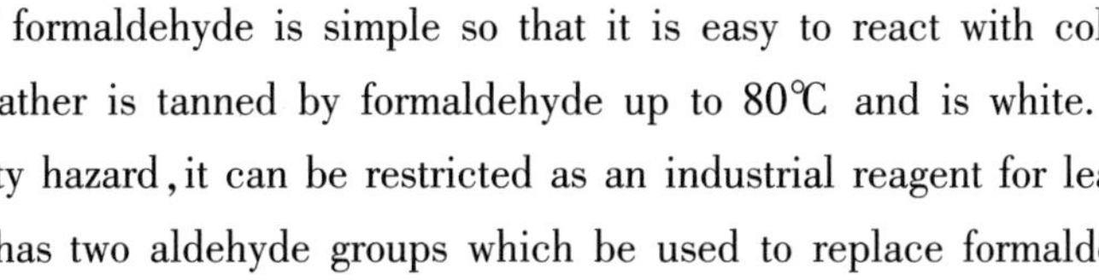

$$R-\overset{O}{\overset{\|}{C}}-H + H_2N-P \rightleftharpoons R-\underset{H}{\overset{O^-}{C}}-\underset{H}{\overset{H}{N^+}}-P \xrightleftharpoons{-H_2O} R-\underset{H}{C}=N-P$$

Figure 3. 26 Reactions between aldehyde and amino groups of protein (P=protein)

The structure of formaldehyde is simple so that it is easy to react with collagen. The shrinkage temperature of the leather is tanned by formaldehyde up to 80℃ and is white. However, because of formaldehyde's toxicity hazard, it can be restricted as an industrial reagent for leather making.

Glutaraldehyde has two aldehyde groups which be used to replace formaldehyde. The shrinkage temperature of the leather tanned by glutaraldehyde is similar to that of formaldehyde. But the color of the leather is yellow and appears orange astime goes on, which has something to do with changes in structure upon reaction with collagen. Therefore glutaraldehyde was modified by formaldehyde, amino acid and other moners (as shown in Figure 3. 27) so that modified glutaraldehyde is obtained and is currently widely used in leather-making.

In addition, aldehydic agents such as oxazolidine, phosphonium salt, etc have good tanning

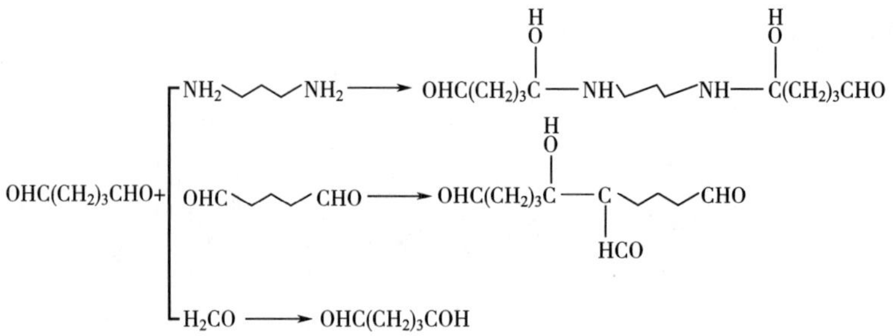

$$OHC(CH_2)_3CHO + \begin{cases} NH_2\sim\sim NH_2 \longrightarrow OHC(CH_2)_3C(OH)H-NH\sim\sim NH-C(OH)H(CH_2)_3CHO \\ OHC\sim\sim CHO \longrightarrow OHC(CH_2)_3C(OH)H-C(HCO)\sim\sim CHO \\ H_2CO \longrightarrow OHC(CH_2)_3COH \end{cases}$$

Figure 3. 27 The modified methods of glutaraldehyde

properties. Although they are distinguished from conventional aldehyde, they undergo some aspect of the typical aldehyde reaction or they exhibit analogous reactions.

Oxazolidine is a compound containing an aliphatic heterocyclic ring which is synthesized by alcoholamine and aldehyde. The ring can be opened to form N - methylol groups. The chemistry reaction between oxazolidine and protein is aldehydic. The monocyclic, bicyclic structure of oxazolidine and the mechanism of reaction with protein is shown in Figure 3. 28 and Figure 3. 29, respectively. The tanning reactions with collagen and the characteristics of leather are similar to those of aldehyde tanning. Oxazolidine has good tanning properties so that the shrinkage temperature of the leather tanned by oxazolidine can obtain the range of between 80℃ and 85℃ which is slightly higher than that of aldehyde tanning. The softness, fullness, sweat resistance, washing resistance, and tear resistance of the finished leather are better than those of the leather with glutaraldehyde.

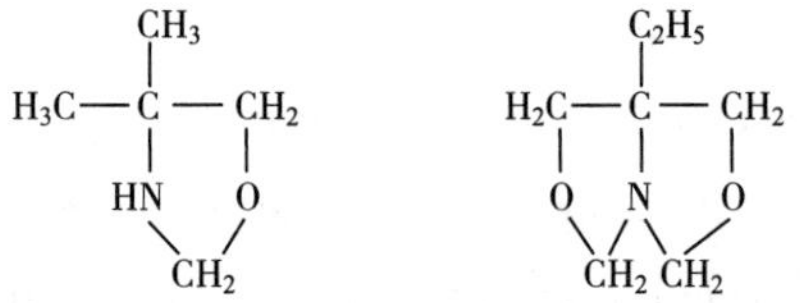

Figure 3. 28 The structure of oxazolidine

$$\text{bicyclic oxazolidine } (C_2H_5) \xrightarrow{H_2O} HOCH_2-C(C_2H_5)(CH_2OH)-N(CH_2OH)_2$$

$$HOCH_2-C(C_2H_5)(CH_2OH)-N(CH_2OH)_2 + 2P-NH_2 \longrightarrow HOCH_2-C(C_2H_5)(CH_2OH)-N(CH_2-NH-P)_2$$

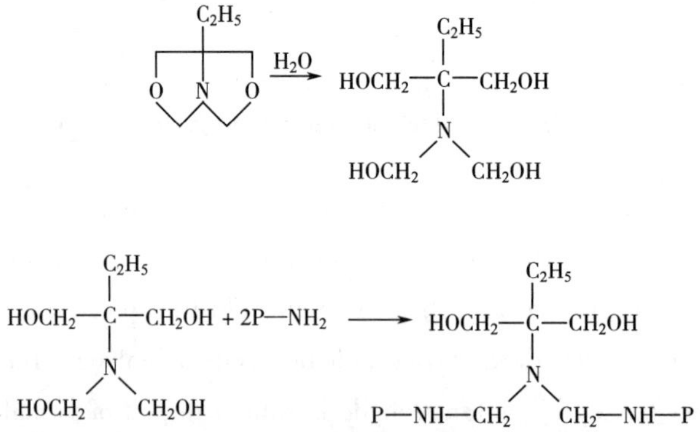

Figure 3. 29 The mechanism of reaction between oxazolidine and collagen

In recent years, tetrakis hydroxymethyl phosphonium sulfate (THPS) and tetrakis hydroxymethyl phosphonium chloride (THPC) have become available commercially which are used as biocides before it acts as tanning agents. The structure of THPS is presented in Figure 3. 30.

$$\left[HOCH_2-\overset{\displaystyle CH_2OH}{\underset{\displaystyle CH_2OH}{\overset{|}{\underset{|}{P^+}}}}-CH_2OH \right]_2 SO_4^{2-}$$

Figure 3. 30 The structure of tetrakis hydroxymethyl phosphonium sulfate

Phosphonium salt can improve the hydrothermal stability of collagen by the cross-linking reaction between hydroxymethyl groups and amino groups of collagen. The shrinkage temperature of the leather tanned by THPS can obtain 74℃ or so. The leather has a strong positive charge which is good for the absorption and binding of anionic chemical materials. But the research found that phosphonium salt can release free formaldehyde.

THPS is a potent tanning agent and is likely to be a useful addition not only to the tanning industry but also to all other sectors that rely on the chemical stabilization of protein by covalent organic means.

3. 4 Dyehouse operation

3-7

Dyehouse operation refers tothe wet processing steps that follow the primary tanning reaction which includes rewetting process & neutralization, retanning, dyeing, and fatliquoring operations. Although each of these operations has a very different purpose, the tanners consider them as a unit because they follow one another without interruption, requiring a total time of about 4 to 6 hours. Based on the requirement of the type, and style of the finished leather, the different retanning agents, neutralizing materials, dyestuff, fat-liquoring agents, and other auxiliaries are chosen, designed and forming practiced formulation so as to obtain desirable leather.

3. 4. 1 Rewetting process and neutralization

According to the requirement of the type and style of the finished leather, the wet blue or wet white is sorted, wrung, split, shaved, or buffed. Then they will be rewetted during the rewetting process whose functions are as follows:

(1) Refilling with free water and good for the penetration of subsequent materials.

(2) To digest the neutral salt of the leather, open up the weak bonds such as hydrogen bonds, electrovalent bonds, etc, and improve the reactive activity of the leather.

(3) To wash away the dirt on the surface of the leather and lighten the color of the surface of the leather.

(4) To adjust the pH of the leather and prepare for the next process.

Therefore the rewetting process plays an important role during the dyehouse operation. Organic acids, surfactants, acrylic resin with low molecular weight, auxiliary aromatic syntan and proteinase are employed in order to obtain the ideal results of re-wetting. Organic acids include formic acid, acetic acid, and oxalic acid which can increase the solubility of metal salts, lighten the color of the leather, and open up the hydrogen bond between collagen and other materials. Surfactants consist of anionic surfactants, nonionic surfactants which can degrease and benefit for the penetration of other materials. During the production of white leather or leather with light color, acrylic resin with low molecular weight is used and can react with or replace parts of the chromium (Ⅲ) on the surface of the leather. Auxiliary aromatic syntan is a non-swelling, and strong acid which can disperse the excessive binding tanning agents on the surface of the leather, replace the metal salts binding with the amino groups of the collagen and from the clean, even buffered interfaces. If the degree of dehydroation of the leather is high or the measure of opening up of collagen fibers is poor, neutral or acid proteinase will be employed to hydrolysis the adhensiving substances among fibers and other proteins and achieve to disperse the collagen fibers and rehydration.

Neutralization process is an important process during dyehouse operation and directly determines the degree of permeability of anionic materials such as dyestuff, organic retanning agents, fat-liquoring agents, etc. The functions of neutralization are as follows.

(1) To remove the part of acid in the leather which is produced because of the hydrosis of Chromium (Ⅲ) salts or further binding between Chromium (Ⅲ) salts and collagen.

(2) To change the charge of the leather by changing the pH of the leather and approaching the isoelectronic point (p*I*) of the leather.

(3) To increase the reaction between Chromium (Ⅲ) salts and collagen.

(4) To weaken the reactive activities of Chromium (Ⅲ) in the leather.

The pH of wet blue is 3.8-4.2 and decreases to 3.6 or so with the extension of the storage time. Therefore the leather protein is cationic and is further accentuated by the mineral retannage under acid conditions. Consequently, negatively charged materials such as vegetable tannins, amino resins, dyestuffs, fat-liquors, etc will readily precipitate on the surface of the leather. Neutralizing can remove the acid in the leather and hence reduce the cationic charge so that there is less reaction with these anionic materials, which then penetrate the leather more uniformly and thoroughly.

During the neutralization operation, sodium formate, sodium acetate, calcium formate, sodium bisulfite, sodium bicarbonate, ammonium bicarbonate, neutralization & retanning agents, sodium thiosulfate, etc are employed to adjust the pH of the leather. According to the state of the leather and the requirement of the next process, the degree of neutralization process depends on the requirements of retanning, dyeing, and fat-liquoring and is controlled by using of different neutralization materials and their dosage. For tight or thick chrome tanned leather, the neutralization pH is controlled to be at the range of 4.5 and 5.4 by using sodium formate, sodium acetate, sodium bicarbonate and neutralizating & retanning agents, which can be measured with pH indicator and bromcresol green indicator by cutting the cross-section of the leather, which can ensure and achieve the better reaction between the leather and other anionic materials because of cationic charge of the leather (the pI of chrome tanned leather is at the range of 6.5 and 6.8). For soft chrome tanned leather such as garment leather, glove leather, furniture leather, etc, the neutralization pH is controlled to be in the range of 5.5 and 6.5 by using sodium formate, sodium bicarbonate, and neutralization & retanning agents, which can be measured with pH indicator and methyl red indicator by cutting the cross - section of the leather. Thus ensure and achieve better penetration, and reaction between the leather and other anionic materials because of weak catioinic charge of the leather. But the neutralization pH of the chrome-tanned leather can't exceed 7.0 so that the grain of the leather become coarse, losing flexibility, crack, and dyeing uniformly. Therefore caustic alkalis are avoided in the neutralizing process.

In addition, the consumption of rewetting agents or neutralizating materials, the float ratio, the temperature of the float, and the reactive time play important roles during the rewetting or neutralizating process. In general, the temperature, the float ratio and the time of the rewetting process is the arrange of 35℃ and 45℃, 1.5 and 2.5, 30min and 60min, respectively. The temperature, the float ratio and the time of the neutralizating process is normal temperature (18-22℃) or 30℃ and 35℃, 1.5 and 2.5, 60min and 120min, respectively.

After the the rewetting or neutralization process, the leather is washed and enter the next process in order to remove neutral salts in the leather. If the neutralized chrome-tanned leather is delayed, the leather will start to generate acid because of the hydrolysis of chromium(Ⅲ) salts. The delayed leather should be neutralized by the neutralizing agents before it starts to enter into next process.

3.4.2 Retanning process

As we know, different tanning agents and their tannage can give different properties to leather. But adopting a kind of tanning agent a tannage is a difficult to meet the requirements of the final leather products. Another tanning process needs to be operated so that it can append or strengthen the fore-tanning properties and give some special properties to leather. Therfore, retanning process is also an important process during dyehouse operation, which can improve the properties of tanning, make up for the defects of the leather, give new and special properties, increase the area of the finished leather, and is honored as the golden touch in modern technology of leather manufacture. The design and result of the retanning formulations directly determine the style, and quality of the finished leather.

In order toproduce the leather with the characteristics of fullness, softness, and resiliency, it is necessary to retan the wet blue with appropriate tanning agents. These tanning agents include inorganic retanning agents (mineral tanning agents), organic re - tanning agents, multi - component retanning agents, etc. They have good retanning, filling properties and give the leather various characteristics and styles. These re-tanning agents are chosen according to the state of the leather, the requirement of the finished leather, the different characteristics of different retanning agents, etc.

3.4.2.1 Retanning with mineral tanning agents

In general, inorganic retanning agents (mineral tanning agents) consist of basechromium (Ⅲ) salts, basic aluminium (Ⅲ) salts, zirconium (Ⅳ) salts, etc.

In order to further improve the shrinkage temperature and increase the chromium (Ⅲ) content of the chrome - tanned leather, even up the color, change the reactivity of the leather, obtain great softness, a pleasant handle, a smooth finely porous grain and favorable coloristic properties, the best-known method is retanning with chrome tanning agents added in proportions of 1%–2% of chromium oxide based on the weight of shaved leather. The shrinkage temperature of the retanned chrome-tanned leather can obtain 120℃ or so, the content of chromium (Ⅲ) of the chrome-tanned leather is evidently improved, and the distribution of chromium (Ⅲ) in the leather is uniform, so that the characteristics of the chrome-tanned leather are more prominent than before and the reaction activities of the leather are stronger than before.

Rechroming is often used if wet blue is purchased from different sources, in an attempt to make the color more uniform between batches of the leather.

But the parts of the hide with a loose structure are not adequately filled in the re-tanning process with base chromium (Ⅲ) salts so that the leather tends to be comparatively empty and its location difference can't be improved. In addition, the suede has an unlevel nap because the fibers of the chrome-tanned leather are soft and tough.

Different from the tanning process, the basicity degree of chromium (Ⅲ) salts in the reranning process is at the range of 35% and 42% which have more molecular weight and form the cross-linking among the collagen fibers. However the retanning operation of chromium (Ⅲ) salts is similar to that of the tanning process.

Basic aluminium (Ⅲ) salts, zirconium (Ⅳ) salt are also used for retanning. The main feature that distinguishes them chemically from chromium (Ⅲ) is that the atomic structure does not allow them to form penetration complexes. This entails that all ligands could form compounds with zirconium or aluminium salts are exchanged much more rapidly than is the case with chrome. Thus both these types of tanning agents are more cationic than chromium salts. The consequence is that zirconium and aluminium tanning agents are precipitated more rapidly in the outer layers of the leather. Thus, there is a specific filling effect within the capillary cavities of the grain and in the reticular layer immediately below it. As a result, the grain becomes denser, smoother, flatter and finer. At the same time, the extensibility of the leather is reduced. The good filling action in the grain layer gives rise to leather that is readily buffable, and a fine, uniform short nap is obtained. For this reason, the main fields of application for these tanning agents are all kinds of buffed leather. Retannage with aluminium or zirconium also improves the structure of the grain and the grain firmness of full grain leathers obtained from spongy raw hides. Obviously, since they are deposited on the surface, they exert less filling action on the interior than chrome tanning agents. The filling action of zirconium tanning agents is somewhat more pronounced than that of aluminium compounds. Therefore, the leather has to be further retanned with tanning agents that exert a filling effect.

Aluminium and zircomium tanning agents are strongly cationic and are liable to hydrolyze. Therefore, they must be used in the acid pH range, and retanned leather can get a high color intensity and billiance after dyeing.

3.4.2.2 Retanning with organic retanning agents

Organic retanning agents include vegetable tannins, syntans, amino-resins, polymer retanning agents, etc.

Vegetable tannins contain hydrolyzable tannins and condensed tannins, have good retanning & filling properties and are widely used in the retanning process of leathers such as shoe upper leather, furniture leather, bag leather, parts of garment leather, etc. They have a mimosa, TARA, quebracho, chestnut, myrobalans, etc. The retanning process of vegetable tannins is shown in Figure 3. 31.

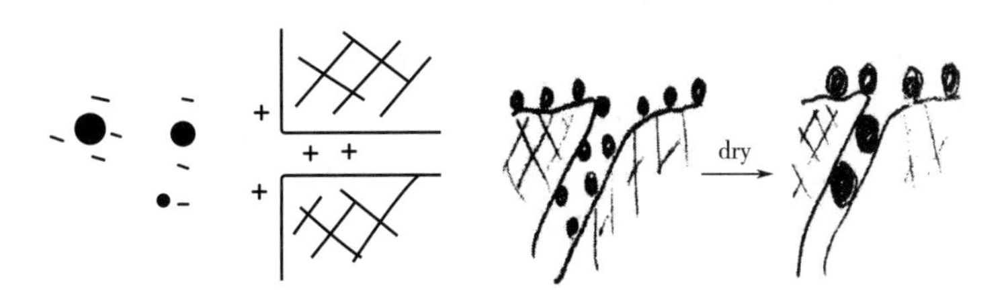

Figure 3. 31 The retanning process of vegetable tannins

The retanned leather has strong anionic and reactivities and is fullness, tightness. Whereas the leather has strong hydrophilic abilities and is not easy to be dyeing. The powers of antioxidation, mold resistance and iron ion (Ⅲ) resistance decrease. The strength, degree of smoothness of the grain of the leather also decrease.

In general, the consumption of vegetable tannins is in the range of 2% and 3% based on the weight of shaved leather in the retanning process. If the specific vegetable feel and filling effect are required, up to 10% of vegetable tannins are used.

A combination with sytans is often used. Syntans act on vegetable tannins in a dispersing way, they improve their diffusion into leather and prevent surface from overtanning. A combination with filling resins has a similar effect.

Generally, the retanning step takes much less time to be accomplished than that of the original chrome tannange. First of all, the leathers are washed and neutralized with mildly alkaline chemicals to adjust both the temperature and pH of the system to the best levels for the particular retanning materials Then the chosen material is introduced into the turning drum and combines with the leather usually within 60 to 120 minutes.

In order to retain the advantanges of vegetable tannins and overcome their shortcoming, syntan is used to replace all or part of vegetable tannins in the retanning process. As previously mentioned, syntans include auxiliary syntans, replacement syntans and comprehensive syntans. Repalcement syntans have a few hydrogen groups and medium or big molecular weight so that they can be used to replace vegetable tannins and give the leather fullness and light color. Auxiliary syntans can be used for dispersing the retanning mateials such as vegetable tannins, replacement syntans, polymers, etc. Comprehensive syntans have different dispersing and tanning properties according to their structure. The tanning operation is similar to that of vegetable tannins.

Amino resins can also be used in the retanning process. They were water−soluble methylol compounds of various nitrogenous bases, particularly dicyandiamide, melamine and urea which contain hydroxymethyl methylol groups and have big molecular weight. The hydroxymethyl methylol groups can react with amino groups, and peptide bonds of collagen by covalent bonds and hydrogen bonds respectively. In addition, they can be precipitated by reducing the pH. Amino resins have selective filling properties so that they can improve the location difference and loose grain of the leather. The color, grain of the leather retanned by amino resin are light and tight, respectively. The leather has strong hydrophilic abilities, poor aging resistance and releases free formaldehyde during the being stored.

The structure of the pores, the break and the handle are much more similar to those of pure chrome leather so that they are very suitable for retanning corrected leather. When they are used for full grain leather, the consumption is restricted to a range of 2% and 3%. The reaction of melamine and for maldehyde is shown in Figure 3. 32.

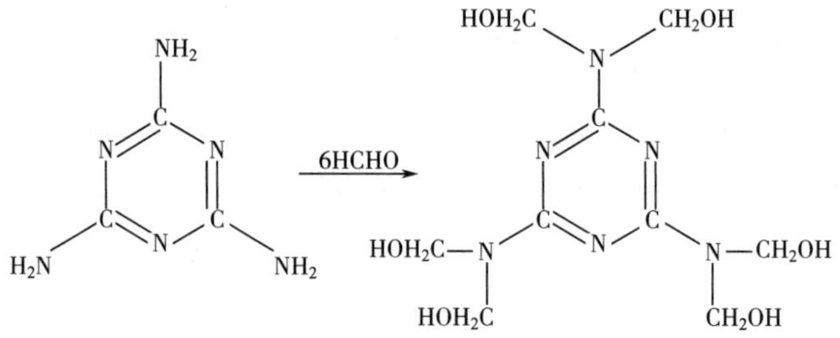

Figure 3. 32 The reaction of melamine and formaldehyde

Polymer retanning agents include polyacrylic acids, styrene-maleic anhydride copolymers with high molecular weight. Polyacrylic acids are synthesized with acrylic acid and other moners by radical copolymerization. Whereas styrene-maleic anhydride copolymers are synthesized with styrene and maleic anhydride by radical co-polymerization (Figure 3. 33). They contain carboxyl groups which have little or no affinity for untanned hides or strongly anionic leathers and form complexes on mineral-tanned or retanned leather.

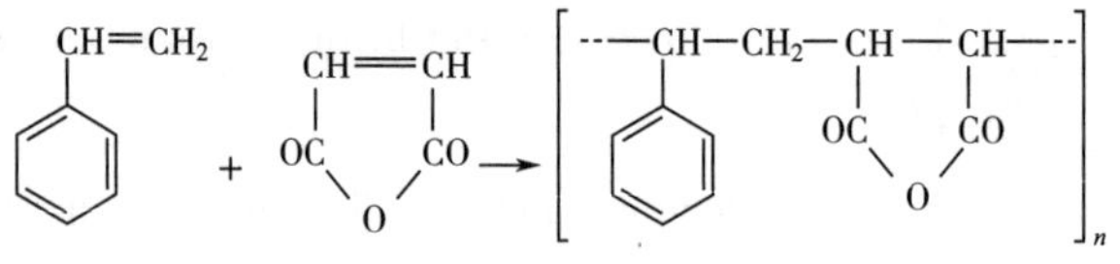

Figure 3. 33 Reaction between styrene and maleicanhydride

Polymer tanning agents exert a selective filling effect on the loose parts, the butts of the leather so that they are very suitable for retanning all kinds of leathers. The leather has fullness, fine grain, good resiliency, rubbery handle and good light fastness. But the leather retanned by polymers can't be dyed in the dyeing process because of stronger negative charge, which is the effect of light color. In addition, they have strong plastic feel when a few polymers are applied in the retanning process. They will become hard under the conditions of high temperature and high pressure.

Aldehyde tanning agents are also used in retanning and give leather good retanning properties.

3.4.2.3 Retanning with multi-component retanning agents

In order to aggregate the advantages of different tanning agents, some tanning agents are studied and produced to a retanning agent with more components such as chromium-aluminium retanning agents, chromium-aluminium retanning agents, syntan containing chromium (Ⅲ), etc. They are used for the retanning of the leather and the tanning operation is similar to that of chromium (Ⅲ) salts.

During the operation, the selection, consumption and sequence of retanning agents and controllment conditions of retanning are very important according to the state of the leather and the requirement of the finished leather. They will form a desirable tanning formulation to meet the requirement.

3.4.3 Dyeing process

Dyeing is one of the more important steps in leather making as it is usually the first property of the leather to be assessed by the consumer or customer. They will make judgements at a glance: color, depth of shade, uniformity. Therefore, it is critical that the science and hence the technology of coloration is well understood. The standard system for measuring color defines the coordinates in the CIELAB color space: a^* is the red-green component, b^* is the blue-yellow component and L^* is the black-white component from which the other parameters can be calculated, such as the hue angle and the chroma. These measurements are useful in comparing color and for color matching. When the dyestuff confers transparent color to leather, the outcome is dependent on some parameters: dyestuff chemistry and mechanism of fixation, relative affinities of the dyestuff and the substrate, nature of the substrate, including its color, illumination of the leather: the perceived color depends on the light source.

3.4.3.1 Dyestuff

During the dyeing process, one of the most important materials is dyestuff which can give different colors by solo or matching two or more kinds of dyestuffs. The modern synthetic dyestuffs industry was initiated by the development of mauveine by Perkin in 1856: the primary markets were textiles, but the leather industry eventually took advantage half a century later, especially when it was realised that bright deep shades could be achieved with the new tannage with chromium (Ⅲ). The dyestuffs include acid dyes, basic dyes, direct dyes, mordant dyes, premetallised dyes, reactive dyes, sulfur dyes, etc. Acid dyes, basic dyes, direct dyes, mordant dyes, premetallised dyes, reactive dyes can be used in the dyeing process. Usually acid dyes, direct dyes are used for the dyeing of the leather. Now, special dyes for leather are exploited and widely used in the dyeing of the leather.

Acid dyesare so called because they are fixed under acid conditions which have a wide range of colors, offering bright deep shades and is the most commonly used in the leather industry, particularly for chrome tanned leather. The structure of acid red and acid orange is shown in Figure 3.34. The molecular weight of acid dyes is relatively small and used for penetrating dyeing, producing level shades. The charge of the dyestuff is anionic so that it has a high affinity for cationic leather. They can react predominantly through an electrostatic reaction between their sulfonate groups of acid dyes and the protonated amino groups of the leather. Simultaneously another reaction is via hydrogen bonding through autochrome groups. Acid dyes are fixed by acidification due to the presence of sulfonate groups and have good fastness properties.

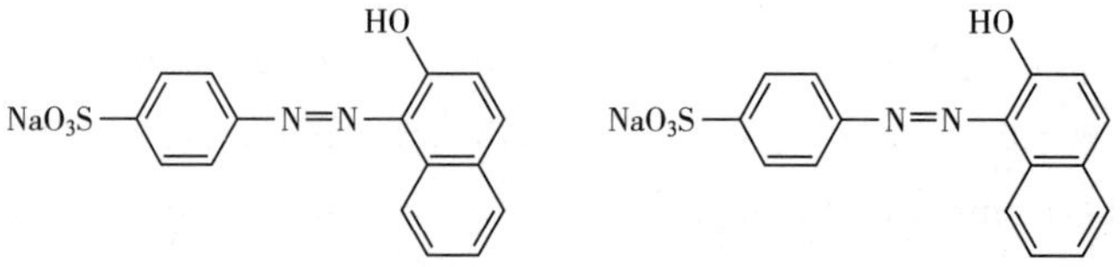

(a) acid red

(b) acid orange

Figure 3. 34 The structure of acid red and acid orange

Direct dyes have the same sort of structural features as acid dyes and have larger molecular weight so that it can be used for surface dyeing with the consequent likelihood of uneven coloring and more direct reaction, not requiring fixation by pH adjustment. The structure of direct con go red is shown in Figure 3. 35. They rely more on hydrogen bonding from a larger number of auxochromes per molecule, similar to the relative astringencies of vegetable tanning agents, and more emphasis on hydrophobic bonding so as to have good fastness properties.

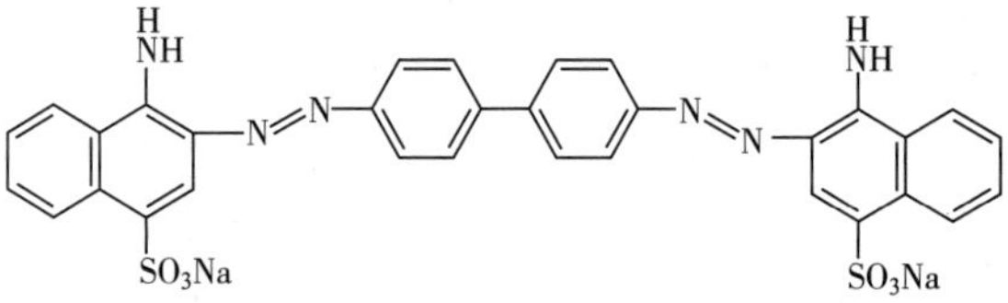

Figure 3. 35 The structure of direct congo red

Basic dyes have strong, brilliant colors such as red, orange, yellow, green, blue, indigo, violet and black. Their structures are essentially the same as the acid dyes and contain the cationically charged amino substituents and anionic groups so that they carry a net positive charge. There fore the dyes have a high affinity for anionic leather, e. g. vegetable tanned, anionic retanned, acid dyed leathers, and are exploited in "sandwich dyeing": acid dye, then basic dye, possibly topped with more acid dye so as to create deep shades with good rub fastness by the electrostatic attraction between the charged species.

However, basic dyes tend to be relatively hydrophobic and often soluble in oils and non-aqueous solvents because they contain fewer solubilizing groups than acid dyes. The dyes have a greater affinity for themselves than for the aqueous solution or the substrate so that they tend to bronze, i. e. produce a metallic sheen. The basic dyes are applied by mixing with acetic acid, then diluting with hot water.

Basic dyes react with collagen fibers by electrostatically (their protonated amino groups and ionized carboxyl groups on collagen), hydrogen bonding and hydrophobic bonding. In addition, the basic dyes have poor light fastness and good perspiration fastness and can be precipitated by hard water and anionic reagents.

Premetallising dyes include pale, dull, and pastel shades, and are obtained to avoid the two-step process of mordanting then dyeing by preparing the complex of dye and metal salt in advance, which is an example of a compact process. 1 : 1 permetallised dyes and 1 : 2 permetallised dyes are shown in Figure 3. 36, Figure 3. 37. They have lower anionic charges than the corresponding anionic uncomplexed dye and good penetrating, leveling properties. The fixation mechanism of 1 : 1 permetallised dyes is the reactions between the metal ion and collagen carboxyl groups by coordinate bonding and other reactions forming electrostatic and hydrogen bonding so as to have good fastness properties. Whereas The fixation mechanism of 1 : 2 permetallised dyes is similar to direct dyes. In general, Premetallising dyes are used for dyeing and finishing premium leathers such as gloving, clothing, suede, nubuck, aniline.

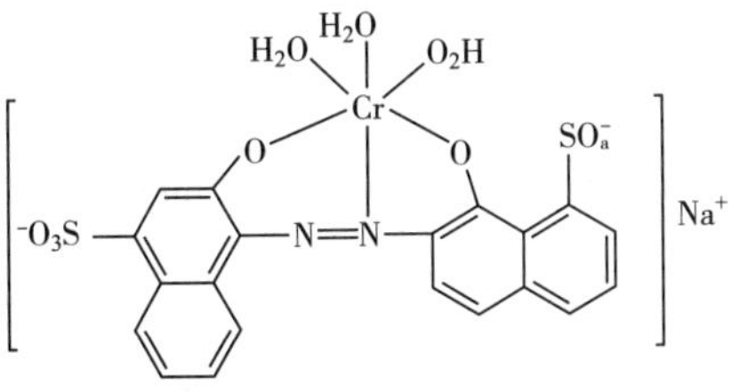

Figure 3.36 1 : 1 chromium (Ⅲ) premetallised dye, acid blue

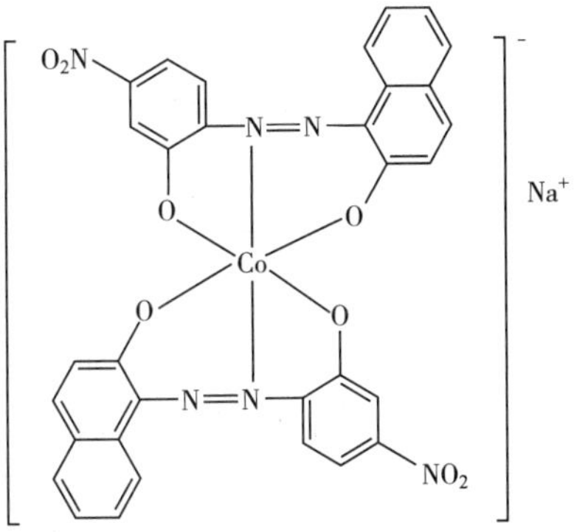

Figure 3.37 1 : 2 premetallised dye, perlon fast violet

Reactive dyes are typically acid dyes that have been covalently bound to a reactive group, capable of reacting covalently with collagen or leather so as to increase the fastness of binding with leather. They are especially useful in applications when the leather may be subjected to washing, dry cleaning or perspiration damage, e.g. clothing and especially gloving leathers. They are expensive and health hazard, due to their reactivity towards organic substrates so as to be restricted.

The structure and components of the leather are different than those of common protein fibers and cellulose fibers. According to the characteristics of the leather, the structure, reaction groups, molecular weight, etc of dyestuff are designed and form the special dyes for leather which combine properties of both acid dyes and direct dyes.

3.4.3.2 Dyeing operation

In general, leathers are dyed either discontinuously in drums or continuously on the machines (through-feed dyeing, spray dyeing or dyeing by roller coating). When they are dyed in drums, it is usually carried out during the processes of retanning and fatliquoring after neutralizing. However, when they are dyed on the machines, it is conducted on the dried crust. The dyes to be used are usually acid and direct dyestuffs as well as metal complex dyestuffs.

The application of leather dyes in drum is a batch process. The dye drums are considerably smaller than those used for chrome tanning. The liquor ratios employed are about 1.5-4.0 based on the weight of the shaved leather. The semi-continuos dyeing of leather on through-feed machine has been in commercial use. This machine makes it possible to dye full grain crust from both sides within a very short time (7-10s) without the leather becoming loose-grained. The main advantage of this machine for tanneries is that the time from the receipt of an order to dispatch of the finished leather is reduced to a minimum. The dyestuff consumption is low. In addition, the method has become more cost-effective: machinery costs have fallen and the formulating vessels used have become smaller, which means that it is no longer necessary to hold such large stocks of chemicals. These represent major savings. Changes in technology have also made it possible to dye leathers of lower shaving substance and even small skin leathers on through-feed dyeing machines. The crust leather to be dyed should be uniform in thickness. They have to be properly retanned and fatliquored. The effect of the retannage on the dyeing properties of the leather is the same as in classical dyeing methods. Leathers with a strong anionic retannage, dye paler than leathers with a less strong retannage. Thus, it is essential for the retannage and fatliquoring to be geared to the through-feed dyeing method if optimum dyeing is to be obtained.

The dyestuffs used for through-feed dyeing should have a sufficiently strong ability to bond to leather within the given time (contact time), thus giving adequate fastness properties. This is not possible with most of the common leather dyestuffs. With direct or acid dyestuffs, there are generally problems with dyestuff migration when the dyed leather is dried, as the dyestuffs do not form a strong enough bond to the leather even under through-feed dyeing conditions and because there is no washing process after dyeing to remove the cutting salts. However, metal complex dyestuffs can form a strong leather-dyestuff bond, resulting in good fastness to water spotting and little or no tendency to migrate on drying. They are therefore particularly suitable for this application. The fastness properties of leathers dyed with these dyestuffs by through-feed method are largely comparable to those of a drum dyeing and meet the requirements for upper leather.

The coloring is accomplished with dyestuffs. The dyestuffs are dissolved in hot water and added through the hollow axle in the rotating drum as soon as the retanning step is completed. The dyestuffs combine with the leather fibers to form an insoluble compound which becomes part of the leather itself. The rate at which the dyestuff exhausts from the color liquor influences the resulting shade, degree of penetration, etc. The faster the exhaustion rate, the greater will be the amount of surface color penetration. In this regard, the tanner once again makes use of pH control in order to help regulate the affinity of the chosen dyestuffs with the leather fibers.

3.4.4 Fat-liquoring process

As the leather dries, the interfibrillary water is removed to allow elements of the fiber structure to come close together and the interactions created by the Maillard reaction occur, which lead to dry, hard leather. The fatliquoring process is essential in order to prevent the reaction from happening and improve the leather quality. Therefore the purpose of the fatliquoring step is to arrest the adhesion of the fibers together during drying. The fibers are lubricated to allow the elements to slide over one another. In the fatliquoring process, oils are introduced into the leather, the individual fibers are uniformly coated with a film of oil as a lubricant so that it reduces internal friction, increases the durability of the leather and prevents fibers from sticking together, thus increasing the softness, stretch, pliability, compressibility and tensile strength.

3.4.4.1 The fatliquoring

The fatliquoring components include the neutral oils, the emulsifying agents and other auxiliaries. The neutral oils originate from animals, vegetables, synthetic or minerals which have the structure of triglyceride or long carbon chain and can give leather softness. The functions of emulsifiers are to emulsify the netural oils and change the netural oils into oil in water emulsions (O/W) consisting of a small drop of oil, surrounded by an emulsifier/detergent/surfactant/tenside. The emulsification methods consist of internal emulsification and outside/external emulsification. The emulsifying agent has a hydrophobic part, which is dissolved in the oil and a charged, hydrophilic part that interacts with the solvent (water), to keep the particle suspended, which is shown in Figure 3.38. The emulsion particles are prevented from coagulating or coalescing because they are held apart by the repulsing effect of the high charges on the surface. Any chemical reaction that reduces the charge on the particle surface will allow them to come together, allowing the neutral oil particles to coalesce. High temperature can drive the particles together, breaking the emulsion. Aims of other auxiliaries is to help the penentration of fat-liquoring agents.

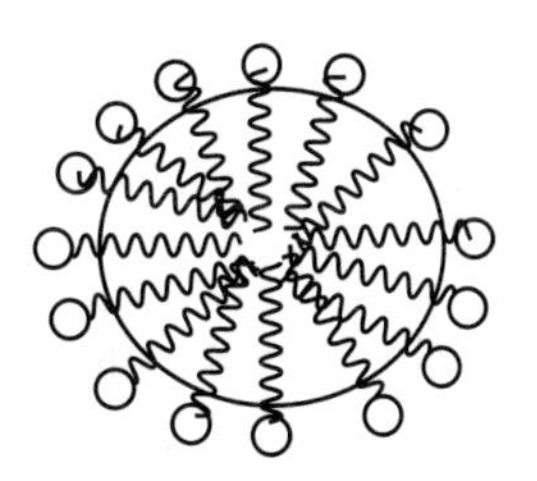

Figure 3.38 Model of an anionic oil-in-water emulsion particle

The netural oils can be directly used, or be used after being modified by concentrated sulfuric acid, phosphorus pentoxide, sodium bisulfite, sodium pyrosulfite, etc. Usually, anionic fatliquors, cationic fat-liquors, nonionic fatliquors and amphoteric fatliquors are used for leather fatliquoring, among them anionic fatliquors widely.

Anionic fatliquors include sulfate fatliquors, sulfited fatliquors, sulfonated oils, phosphate oils, etc. Acted sulfated fatliquors, sulfited fatliquors as examples, and their preparation is illustrated.

For sulfated fatliquors, the oil must be unsaturated, with a minimum iodine value of 70: the iodine value is defined as the number of grams of iodine absorbed by 100 grams of oil or fat. Oils that have been used in this regard are castor, neatsfoot, soya, groundnut, cod. 10%-20% Concentrated sulfuric acid on the weight of the unsaturated oil is added slowly to the oil with constant stirring. The temperature of the exothermic reaction must be controlled to be below 28℃, otherwise the oil can char, causing darkening, and the triglyceride oil may be hydrolyzed to release free fatty acids. The concentrated sulfuric acid is also replaced by the mixture of sulfuric acid and phosphoric acid. Then excess free acid is removed by washing the partially sulfate oil with brine, which also separates the oil fraction from the aqueous fraction. Alternatively, sodium sulfate, ammonium chloride or sulfate could be used. Some hydrolysis of bound sulfate may take place. Lastly, free acid groups are neutralized with alkali in order to increase the hydrophilic properties. Consequently, free sulfuric acid is converted into salts. The fatliquors may therefore be designated low, medium or high sulfated, that is bonding SO_3 may range from 2%, considered a low level, to 3%-4%, considered medium level, to 6%-8%. As the level of sulfation increases, the fatliquors' anionic charge increases, lubricating effect decreases, emulsion particle size decreases, stability of the emulsion to coagulation by acid or metal salts increases, etc. The anionic charge of the fat-liquoring agents increases so as to have a greater affinity for cationic leather. The lubricating effect of the fat-liquoring agents decreases due to the lower concentration of neutral oil. The emulsion particle size of the fatliquoring agents decreases ultimately to the point of forming a microemulsion (<5nm) or even actually dissolving in water. When bonding SO_3 is high levels, the oil functions more like a wetting agent than a lubricant, hence the leather becomes more hydrophilic.

Therefore the low level of sulfated fatliquoring agents are used for drum fatliquoring of vegetable tanned leather because their emulsion has low stability under the conditions of aid or metal salts and can lubricate the surfaces of the leather. The medium level of sulfated fatliquoring agents is used for surface neutralised chrome leather because of the stability of the emulsion and greater penetration. However, the high level of sulfated fat-liquoring agents are used for assisting the penetration of other materials.

For sulfited fatliquors, the is unsaturated oils are required which include cod oil, neatsfoot oil, etc. They are firstly sulfited/oxiadated by air is blown through a mixture of oil and sodium bisulfite solution, with stirring at 60-80℃. Hydrogen peroxide may be used instead of air. Then washing with brine removes excess sodium bisulfite. The fat-liquors have good emulsion stability and better penetration properties so as to make the leather softer and fuller. Therefore the sulfited oils are used for all leathers.

3.4.4.2 Mechanism of fat-liquoring process

The mechanism of fatliquoring can be expressed in simple terms as follows: the neutral oil is transported into the pelt as an oil-in-water emulsion. The emulsifying agent interacts with the leather, reducing or eliminating its emulsifying power. The neutral oil is deposited over the fiber structure-the level of the hierarchy of structure depending on the degree of penetration. The water is removed by drying, allowing the neutral oil to flow over the fiber structure.

3.4.4.3 Fat-liquoring operation

The fat-liquoring results depend on the state of the leather, the quality of the fat-liquoring agents and the fat-liquoring operation. The state of the leather includes the charge of the leather and the degree of open-up of the collagen fibers.

The charge of the leather is related to its tannages, different retanning agents adopted and the degree of neutralization. The thickness of the leather, the degree of the tightness of weaving of collagen fibers, differences in location, the state of the grain of the leather, etc affect the fat-liquoring process.

The quality of the fat-liquoring agents include the kinds of the fat-liquoring agent and its modified methods, the content of active constituents of the fat-liquoring agent and its consumption.

The fat-liquoring agents based on animal oil give leather good softness, fullness and lubricity. Cod oil, neatsfoot oil can give leather good softness and stronger filling properties. Sulfited fish fat-liquoring agents have good penetration properties and the leather is very soft. Lanolin fat-liquoring agents can give good softness, grease touch and silky glossing. The fat-liquoring agents based on vegetable oil have good penetration properties, but the leather fat-liquored has an arid handle and bad grease touch. Although mineral oils give the leather good softness and lubrication and can't produce fatty spew, it can be easy to transfer from the leather because of having a weak affinity for the leather. Synthetic fat-liquoring agents have good light resistance, heat resistance, resistance to electrolytes, filling & penetrating properties and non-producing fatty spew. In addition, the fat-liquoring agents modified by different modifying methods have different properties for the same oil, which are relative to the charge, particle size and the stability of emulsion, etc.

During the fat-liquoring operation, the float ratio, pH, temperature of the float, the consumption of fat-liquoring agents, the reactive time between fat-liquoring agents and the leather and the mechanical actions are important factors affecting the results of fat-liquoring. Moreover, the fat-liquoring operation can be carried out by many steps. They can be implemented in pretanning, retanning, neutralizing process and can also be used for the fat-liquoring of the grain of the leather and oiling off on the parts of the leather. Thus, distributed fat-liquoring is good for improving the absorption of the fat-liquoring agents, uniform distribution of the fat-liquoring agents in the leather and good softness.

3.5 Drying and softening operation

3-8

The drying and softening process of the crust is one of the most important processes during the leather manufacture, which includes the whole processing from drying of wet leather to the finished leather. After the leather is retanned, dyed and fat-liquored, etc, the leathers have physical and chemical properties which will be shown by drying and finishing process of the crust.

3.5.1 The drying process

The main function of drying process of the crust is to remove the surplus water so as to meet the need of the content of water during the subsequent process and the finished leather. During the drying process, the tanning materials, dyestuff and fat-liquoring agents, etc can further react with the collagen so that the tanning effect of the tanning agents is strengthened and the chemical stability of the leather will be further improved. The fiber structure of the leather will be set during the drying process. Therefore the drying process plays an important role in determining the handle, softness, area and thickness.

3.5.1.1 Forms of water in leather and the stages of drying process

After the leathers are set out, the content of the water in the wet leather is 50%-60%. Based on the combing form between the water and the leather, the water in collagen can be divided into three main groups: structural water, bound water and bulk water. Bulk water has a liquid-like character and can form ice crystals at 0℃, which are adsorbed on the surface of the leather and are filled among the fibers. Bound water exhibits a structure between solid and liquid which is condensed in the capillary, firmly combined with the leather and does not freeze at 0℃. Structural water molecules are part of the fiber structure and behave like a solid which can combine with the polar groups of the leather by hydrogen bond. Therefore the removal of the water in the leather can affect the properties of the finished leather.

The drying process includes three stages: preheating stage, constant rate drying stage and deceleration drying stage. The time of preheating stage is short so that the temperature of the wet leather arrives the wet-bulb temperature of the air under the condition of constant temperature and relativehumidity of ambient air around the leather. In the constant rate drying stage, there may be sufficient bulk water on the surface of leather for it to act like a liquid water surface. The rate of evaporation is constant and is proportional to the surface area of leather, mass transfer coefficient and the difference between the vapor pressure of the water at the surface temperature and the partial pressure of the water vapor in the air. The rate of drying is dependent on the temperature and the relative humidity of the air. In theory, fast drying conditions can be used during this phase of drying as long as the losses of the bound and structural water molecules are prevented. After this constant rate of drying period is completed, the rate of drying is controlled by the diffusion of moisture/vapor from the higher concentration in the center of the leather to the lower concentration at the surface. During this period the rate of drying is entirely dependent on the ease of diffusion of water vapor through the structure. The factors that affect the diffusion rate will control the drying rate. When leather is dried at a high temperature, the stiffness is generally found to be high. It is commonly believed that stiffness has a high dependency on the drying temperature. In the deceleration drying stage, capillary water is removed from the surface of the leather and evaporated so as to be easily shrunken of the grain. The drying process can produce softer leather. This confirms that the water-leather relationship during drying plays an important part in determining the softness properties of leather.

3.5.1.2 Drying methods

In general, the drying methods include hanging drying, tacking drying, paste drying, toggle drying, vacuum drying, etc. Different drying methods can give the leather different properties.

Hanging drying is the simplest method of drying leather. The leather is hung in the drying room or drying tunnel and dried by the natural passage of air around it. The drying rate depends on the temperature, the relative humidity of the air around the leather and the air velocity. Hanging drying has the advantages of low capital investment and can obtain full, soft, large elongation leather with clear grain. However, the leather has a coarse grain, uneven surface and lower area yield. The drying methods are suitable for the drying of large stretchy and soft leather such as garment leather, glove leather, etc.

During the pasting drying, the set-out leather is pasted onto the stainless steel plates heated with hot water and the grain surface to the plates. Pasting drying can give excellent area yield and smooth surface which is used for all kinds of leather.

During the toggle drying (Figure 3.39), the toggling unit consists of a number of perforated plates placed in a dryer having controlled temperature and humidity. The tanned leather is stretched and held in the place by a number of toggles that hook into the perforated plates. The segments of the perforated plates are pulled apart mechanically. Compared with hanging drying, 10% more area can be gained.

During the vacuum drying, the set-out leather is spread out, and grain down on a smooth steel plate. Heat is applied to the plate by a heat exchanger under the plate. The plate with the leather can be completely covered with an air-tight hood with a felt mat or fine-mesh screen fabric on the inside. When the hood has been placed on, the air is sucked out to create a vacuum atmosphere. The

Figure 3.39 Toggle drying

water in the leather starts to boil at 45℃ or so and is evaporated until the water content of the leather has dropped to 30%−35%. Drying may be completed in 2−3 minutes. The temperature and the time can be adjusted according to the leather. By vacuum drying, the leather becomes smooth and the area yield is better than that of hanging drying. Vacuum drying is used for shoe upper leather.

Therefore different drying methods have their own characteristics. In general, two or more drying methods are adopted during the drying process according to the requirement of the different leather. Take shoe upper leather as an example, in order to keep the flat grain, fullness, handle, area yield, etc, the set-out leather firstly are dried by vacuum drying, then hanging drying is adopted.

3.5.2 The softening process

During the drying process, the leather has a feeling of stiffness because the leather fibers tend to adhere together. In order to make the leather soft and suitable for various needs, some mechanical operations will be carried out which include conditioning, staking, milling, buffing, polishing or glazing, ironing, etc.

After the leather is dried, the water content of the dried leather is 14%−18%. The fiber structure of the stiff leather is easy to be destroyed when some mechanical operations are working. Conditioning is very important to control the correct moisture content of the leather. Therefore it is necessary to moisten the dried leather by spraying the water on the surface or the other side of the leather or covering it with damp sawdust (30% moisture content). Then the leathers have been piled for 18−24 hours so that the moisture content to be aimed at 28%−30%.

Staking operation is one of the mechanical operations that can make the leather soft. The staking machine and its operating principle are shown in Figure 3.40, Figure 3.41. During the staking operation, the leather is bent and stretched to loosen the fiber bundles which stick together during the drying process. Vibration staking is commonly used for staking operations. The leather is conveyed through the system by laying it between upper and lower elastic conveyor bands. The speed of throughput can be controlled by the speed of these conveyors. The pressure and working action of the tool can be adjusted to suit different leather.

Figure 3. 40 Staking machine

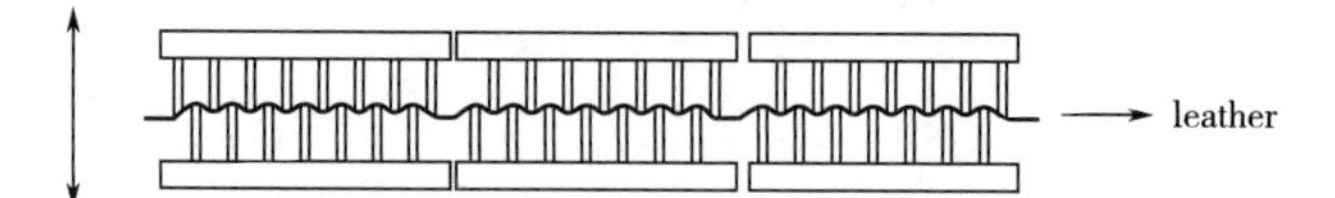

Figure 3. 41 The operating principle of staking machine

Milling operation is also an important mechanical operation which can be carried out in the drum (Figure 3. 42). By controlling the speed and time of rotation, the fiber bundles are loosened and the leather becomes soft. In order to further improve the effect of milling, it can be added into the milling drum that some solid substances such as hard rubber balls, gleditsia seeds, etc and moderate auxiliaries such as feel improvers, softening agents, fat-liquoring agents, wax, water-proof agents, etc. Milled leather has good softness and a beautiful decorative pattern on the surface of the leather. In addition, milling operations with longer time and stronger mechanical actions can obtain the leather with the effect of natural milling. Pattern shape on the surface of the leather is related to the retanning effect and the moisture of the leather.

Figure 3. 42 Milling drum

In order to save the time of milling operation, staking operation can be combined with milling operation.

Buffing operation, polishing or glazing operation, ironing operation, etc are also some of softening processes. Buffing operation can eliminate or reduce minor disability or defects on the surface of the leather, obtain fine, uniform grain and prepare for the subsequent finishing process. Nubuck leather can be obtained by buffing the grain of the leather with high quality. Whereas the uniform piles can be achieved by buffing the Swede leather (Figure 3.43, Figure 3.44).

Figure 3.43 Through feed buffing machine

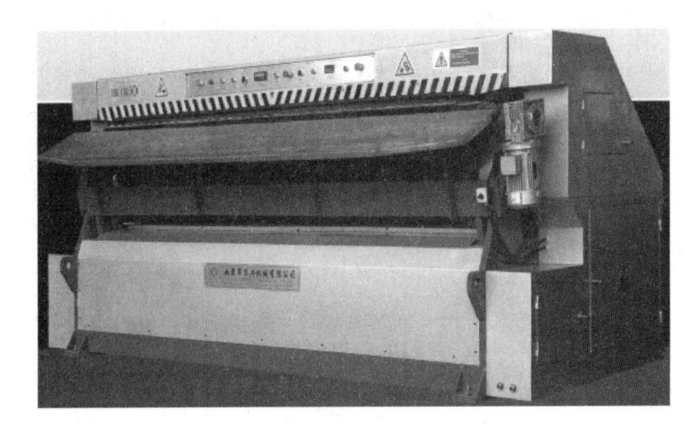

Figure 3.44 Polishing machine

Polishing or glazing and ironing operations are also one of the most important softening processes. By exerting the rolling, extruding, fricative and other mechanical forces on the surface of the leather with the polishing rollers (agate or flock felt), leathers with smooth, plain, tidy, soft and comfortable grain are obtained. At the same time, it can repair slight flaws on the surface of the leather so as to give upgraded leather products. The fine, smooth, tidy leather is achieved by the ironing operation with certain temperature, pressure and velocity. In addition, the ironing operation can improve the gloss of the leather, and reduce slight flaws on the grain of the leather, and improve the water-proof, abrasive properties and tensile strength. In order to achieve the ideal effect of the ironing operation, the temperature, pressure, time, and conveying speed of the roller can be adjusted. Through feed rotary ironing maching is shown in Figure 3.45.

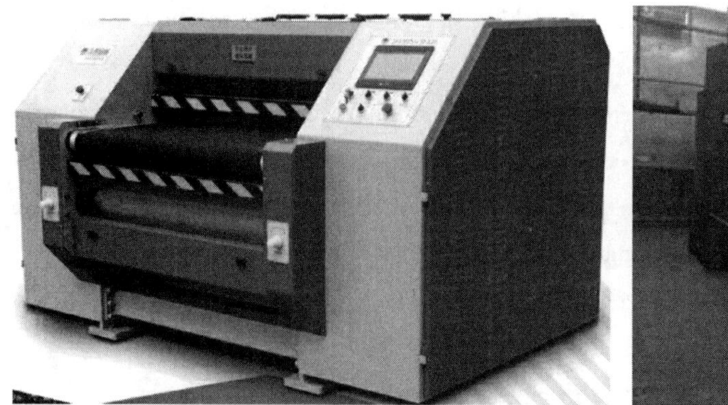 

Figure 3.45 Through feed rotary ironing machine

Therefore, the leather becomes soft and is suitable for being finished or directly being used by the drying process and mechanical operations.

3.6 Finishing

3-9

Finishing operation refers to the further processing of the leather which is recognized as the makeup of the leather. Finishing operation can give a beautiful, level appearance for the finished leather to meet the needs of consumers on color, handle, surface effect, gloss, etc. The finished leather has a smooth and fine surface, temperate gloss, bright and level color, comfortable handle, etc. Finishing operation can also improve the chemical and physical properties of the surface of the finished leather such as heat resistance, cold resistance, water resistance, solvent resistance, and wet/dry rub resistance, etc. Finishing operation can increase the operation rate of the leather and improve the grade of the leather by covering the damage, defects of the leather, then give different kinds of styles such as aniline leather, patent leather, two-tone leather, embossed leather, leather with pull up effect, etc (Figure 3.46, Figure 3.47).

Figure 3.46 The cartoon of finishing operation

(a) Patent leather

(b) Embossed leather

Figure 3. 47　The finished leather

According to the state of the grain of the leather, the finishing operations include the finishing of full grain leather, corrected grain leather and suede leather. In addition, the application of the finishing methods includes brushing, padding, spraying, roller, curtain according to the finishing equipment. Finishing process can be achieved by finishing materials, rational finishing formulations and related equipment.

3. 6. 1　Finishing materials

3-10

Finishing Materials are mixtures which usually consist of film-forming materials, pigment, disperse medium and auxiliaries. Each component plays an important role in the mixture. By formulating these components, various finishes can be obtained so as to be suitable for different leathers.

3. 6. 1. 1　Film-forming materials

Film-forming materials can be used as binding media for non-forming materials such as pigments, wax, fillers, i. e, and can form protective coatings, plain finishes or as fillers and impregnants for soft loose-grained leather. In addition, they can also control the gloss and handle of finishing coatings. According to the degree of softness or hardness of the film, the type of film-forming materials include frizzed soft, soft, middle soft, middle hard, hard. They are usually acrylate resins, butadiene resins, polyurethane resins, proteins, cellulose, compact resins, etc.

- Acrylate resins emulsion

Acrylate resins emulsion is synthesized by the radical co-polymerisation reaction with acrylate, acrylic acid or methacrylic acid which has good stability, excellent properties of film-forming and blending properties with pigments, auxiliaries and other film forming materials. They can give plastic transparent films which own good light resistance, adhesion, embossing properties, solvent resistance and gloss. Based on the different standards of the classification, acrylate resins emulsion have different types. They mainly have filling, self cross-linking, anionic or cationic types and different softness and gloss.

- Butadiene resins

Butadiene resins are polymers based on butadiene monomers or copolymers based on butadiene monomers and acrylate. The resins have good film form properties, excellent covering properties and good physical & mechanical strength. But the film has bad light resistance because of the carbon-

carbon double bond in the structure of the polymers so that they are rarely used in the finishing of the leather.

- Polyurethane

Polyurethane is a very large group of resins which are obtained by condensation polymerization between polyisocyanate and polyester polyol or polyether polyol. The products can form tough resistant films which are used to finish a variety of different types of leathers. Compared with the properties of films of acrylate resins, the film of polyurethane owns excellent adhesive properties, heat and cold resistance, physical and mechanical properties, high gloss, light resistance and shear – resistant properties. They can be classified into solvent lacquers and dispersions. Solvent lacquers are soluble in organic solvents and are usually used for patent and wet look finishes. Dispersions can be used as base coats and binders in leather finishing to produce finishes with excellent adhesion and high scuff resistance.

- Proteins

Proteins include casein, albumen, gelatin, and synthetic polyamides which can be used as film-forming materials. The film has good plate release property, can improve the pilling properties and natural, comfortable handle. In addition, they give heat and organic solvent resistant films, poor wet rubbing resistance because they are not easily made insoluble in water. They are employed for glazed finishes on calf, goats, sides, and some specialty leathers.

- Nitrocellulose lacquers

Nitrocellulose lacquers are obtained by nitration between cellulose and mixed acids including nitric acid and sulphuric acid (Figure 3.48). Cellulose acetate lacquers are obtained by nitration between cellulose and acetic anhydride under the condition of sulphuric acid. The degree of esterification directly affects the properties of nitrocellulose lacquers and cellulose acetate laqcquers. In general, the film has good gloss, comfortable handle and water-reistance properties so that it can be used for the finishing of top coating as lacquers and in isolating layers for isolating among layers and preventing the coating from sticking board.

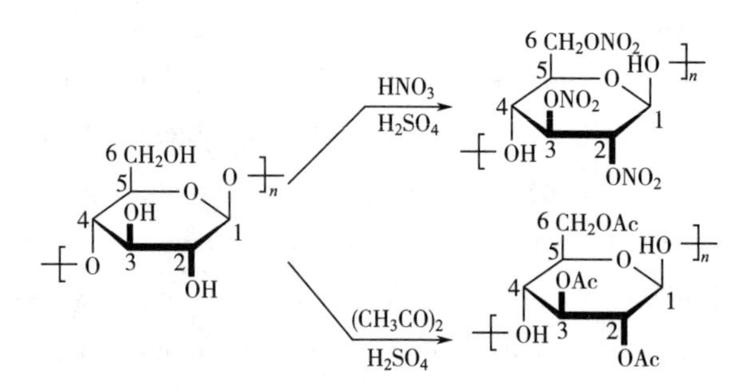

Figure 3.48 Reaction and its structure of nitrocellulose lacquers

3.6.1.2 Pigments

Pigments can give the leather different colors which are insoluble in water or solvents and opaque. They include inorganic or organic pigment. Inorganic pigments are prepared by grinding natural, colored rocks or ores to fine powders and then mixed in a mill with water and a wetting agent or a binder such as casein solution or other synthetic polymer dispersions. The structure of organic pigments is similar to those of the dyestuff and gives bright color to the leather. In addition, the pigment is divided into the one including casein and the one including no casein according to the kinds of binders. The pigment including casein has good filling, forming films, hot-resistance properties and gives good plate release properties. However, it gives leather good covering power so as to be commonly used in paint. Several pigments with different colors are blended in order to obtain the aimed color, which is followed as Table 3.4.

Table 3.4 Combination results of two or three different pigments

Pigment Ⅰ	Pigment Ⅱ	Pigment Ⅲ	Pigment with aimed color
Red	Yellow	—	Orange
Red	Blue	—	Purple
Yellow	Blue	—	Green
Purple	Yellow	—	Black
Blue	Orange	—	Black
Purple	Orange	—	Black brown
Red	Green	—	Yellow
Blue	Green	—	Military green
Red	Blue	—	Fuchsia
Red	Yellow	Blue	Black

3.6.1.3 Auxiliaries

Auxiliaries are one of the most important ingredients in the paint which include penetrating agents, fillers, wax emulsions, matting agents, cross-linkers, levelers, hand modifiers, etc. They can modify the appearance, performance, and handle or feel of the leather by assisting other materials such as binders, pigments, etc. Penetrating agents can adjust the degree of penetration of finishing materials and often be used in base coats. Fillers are non-forming film materials which is a composite of silicon dioxide, kaolin, protein, wax emulsion and polymer compounds. They have good covering and filling properties so that the film containing the filler is opaque. Wax emulsion is the emulsion of beeswax, carnauba wax, microcrystalline wax, polyethylene wax or other waxes which can provide the waxy feel and plate release for the coating. Matting agents are composite among ludox with homogeneous size particles, stearate and its derivatives, acrylamide polymer, titanium dioxide and wax emulsion. When A matting agent is used in the base coating, it can decrease the intensity of refracted light originated from the same direction of coarse surfaces, which is shown in Figure 3.49.

Linear macro-moleculars can be converted into network polymers by the cross-linking of cross-linkers, which is shown in Figure 3.50. Therefore cross-linker can improve the fastness, water resistance, solvent resistance, rub resistance, and heat resistance of the coating if it is used in the paint

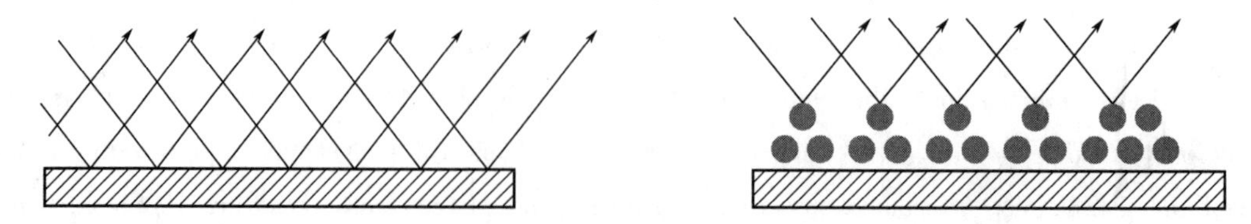

Figure 3.49 Schematic diagram of matting agents

or coating. Cross-linkers include polycarbodiimines, epoxides, isocyanates, aziridines, polyureas. They can react with different binders and obtain the corresponding cross-linking results and the properties of coating (Figure 3.50).

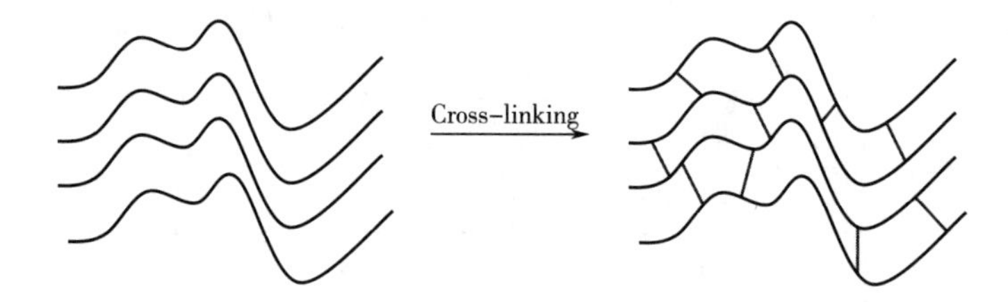

Figure 3.50 Schematic phase diagram of cross-linking

Hand modifiers, whose main ingredients areorganic silicon compounds can give a desire hand such as silk, smooth or other feel for the coating. Levelers can help to adjust the velocity forming films of binders so as to form the smooth and uniform film.

3.6.1.4 Dispersion medium

In general, leather can be finished with either aqueous finishes or organic solvent finishes. In the aqueous finishes, water is used as a medium to disperse film-forming materials, pigments, etc. In organic solvent finishes, organic solvent is used. These dispersion mediums are also employed to control the viscosity of the finish and thickness of finish films on the leather. Compared with the organic solvent, water is cheaper and environmentally friendly. Nowadays, in order to reduce the content of VOC (volatile organic compounds) in the air, water is employed in almost all finishes from base coat to topcoat. Therefore water-based finishing is held in esteem in the leather industry and its market.

3.6.2 Mechanism of forming film of film-forming materials

During the finishing process of the leather, film-forming materials are the one of important components of the finishing materials. It can improve the properties of dry/wet rubbing resistance, gloss, handle, etc of the leather. The quality of the film formation will affect the performance of the coating. The common film-forming materials include solvent type and lotion type ones. They have different mechanism of forming film.

3-11

3.6.2.1 Mechanism of forming film of solvent type filming materials

Solvent type film-forming materials are homogeneous system with polymer and organic solvents. Polymers act as dispersion phases and organic solvent acts as continuous phases. During the film-forming process of the solvent type film-forming agent, the compact, highly bright film is formed because of the evaporation of volatile components of organic solvent and diluent. Therefore the evaporation rate of volatile components of organic solvent and diluent is an important factor affecting the rate of film-forming. However, the evaporation rate is related to the vapor pressure of volatile components. The higher the vapor pressure, the faster the evaporation.

The evaporation process of solvent and diluent from the film-forming materials can be divided into three stages:

The first stage: when it is applied on the surface of the leather as a film agent solution, a large amount of diluent and solvent in the liquid film begin to diffuse upward from the lower layer of the film, the concentration on the film surface increases, and a layer of saturated vapor of solvent gradually forms and begins to evaporate.

The second stage: it first forms a very thin film on the surface of the film-forming agent, that is, viscous gel, and then gradually becomes dry gel, which increases with the thinning of the liquid film. In this process, the evaporation of solvent and diluent needs to overcome the diffusion resistance of the whole film and the resistance of the solidified film surface.

The third stage: the solvent most closely bound to the film-forming agent starts to evaporate, and the film shrinks to a stable state.

The physical and mechanical properties of leather coating are directly affected by the film-forming agent. There are many factors affecting the film formation, including the factors of the film-forming agent itself, the diluent and other factors (such as relative humidity, temperature, coating thickness, etc.).

However, the film formed by solvent based film former is uniform, dense, transparent and glossy, but the permeability and water vapor permeability of the film are not good, especially for thick films. In order to maintain the hygienic property of leather as much as possible, a thin coating is also an effective measure.

The volatilization of organic solvent during the film-forming process of solvent based film former will pollute the environment. Therefore, in order to meet the requirements of environmental protection and comfortable wearing, more and more leather products use water-based finishing.

3.6.2.2 Mechanism of forming film of lotion type filming materials

Lotion type filming materials are a heterogeneous system with polymer and water. Polymers act as dispersion phases and water acts as continuous phases. The film-forming process of lotion type film former is more complicated than that of solvent type filming materials. During the film-forming process, lotion particles slowly approach and are bonded, fused with each other so as to finally form a continuous film as the water in the lotion type filming materials volatilizes. The operation and film-forming process of the lotion type filming materials can be divided into four stages which is shown in the Figure 3.51.

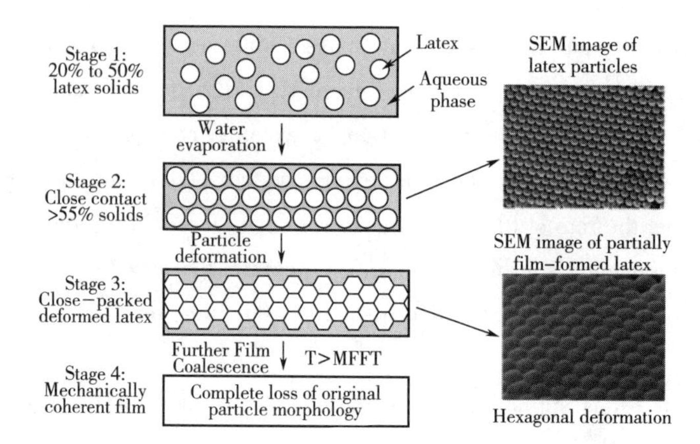

Figure 3.51 Film-forming process of the lotion type filming materials

The first stage is that lotion type filming materials whose contents in the range of between 20% and 50% are coated on the surface of the leather. Then the second stage is that the water in the lotion type filming materials is evaporated and absorbed by the the leather. After the film-forming materials are coated on the surface of the leather, the water volatilizes and is absorbed by the leather. with the gradual decrease of water in the film-forming materials, the distance between the polymer particles is reduced and gradually approach each other, and finally the particles are connected and form a directional arrangement to a certain extent. Thus the contents of lotion type filming materials on the surface of the leather is greater than 55%. The third stage is the fusion process of the polymer particles. As the water in the lotion type filming materials continues to volatilize, the protective layer adsorbed on the surface of the polymer particles is destroyed and the exposed particles close contact with each other and their gaps gradually become smaller. When the gap is small to the capillary diameter, the capillary pressure is higher than the anti-deformation force of polymer particles due to the role of the capillary, and the particles are deformed, agglomerated and fuse into a continuous film. The fourth stage is the diffusion process. Because the emulsion layer is destroyed, the polymer particles interact with each other and form a smooth film by self-adhesion. The emulsifier is dissolved in the polymer or partially absorbed by the leather.

There are many factors affecting the film formation of lotion type film former. It mainly includes the film-forming agent itself, green leather and operating factors (such as the influence of relative humidity of green leather, coating thickness and component materials).

3.6.3 Design of the finishing formulation and its properties

The finishing formulations are very important for the properties of finished leathers. According to the state of the leather and the needs of the finished leather, the finishing formulations can be designed. In general, the coatings include impregnating coating, base coating, middle coating, top coating, touch coating, etc. Different coatings own different finishing formulation. Except for top coating and touch coating, the finishing formulation of other coating is made up of pigment, filler, binders, matting agents and other auxiliaries. Different components can give different functions in the formulation. The first coats applied are the softest and have highly pigmented. The middle and top coats tend to be harder and contain a minimum of coloring matter. The aim of the general system is to keep softness and grain break of the crust. As shown in Figure 3.52, the more the finishing coatings, the lower the grade of the finished leather.

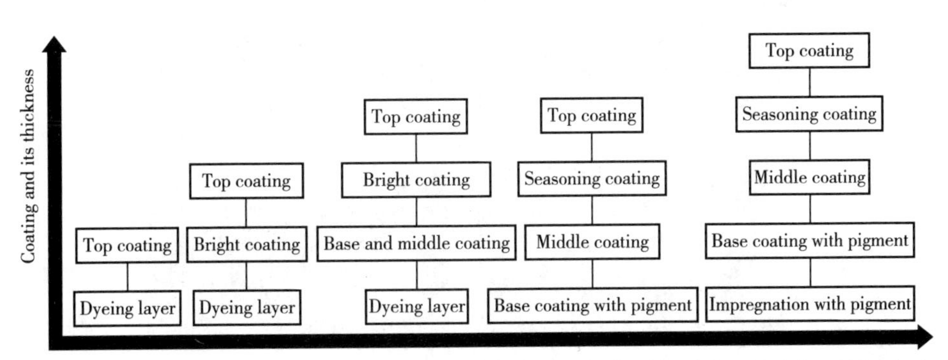

Figure 3.52 Relations between the kinds, number of finishing layers and grade of the finished leather

Based on the conditions of the crusts or economic reasons, there is often no clear difference between these various coats and amalgamation of various coats by using general all-purpose film-forming materials is becoming widespread. According to the requirement of the normal crusts, the finishing formulation can be illustrated and designed, that is, covering-defect coat, spraying dyeing, impregnation coats, base coats, middle coats, seasoning coats and top finishes.

3.6.3.1 Covering-defect coat

In general, most of the crusts have different degrees of defects on the surface (grain) of the crust such as scratches, tick-marks, scars, pin-holes, buffed grain, etc. Some kinds of defects on the surface of the crust are shown in Figure 3.53. These defects affect the grade of the finished leather.

Therefore, in order to cover or fill up the defects, improve the level of evenness in surface appearance, modify the color, gloss level of the finishing layer and increase cutting yield of the finished leather, covering-defect coat is widely used in the finishing of most leathers including imitation full grain leather, corrected leather, suede leather, etc.

The components of the covering-defect coat include stucco, pigment, resin, and water. Recommended usages are followed as:

A: Stucco 1,000

B: Stucco 950 Pigment 50

C: Stucco 900 Pigment 50 water 50

The operation can be achieved by injection or spot-spatula filling, hand-spatula, padding, spraying and compact methods.

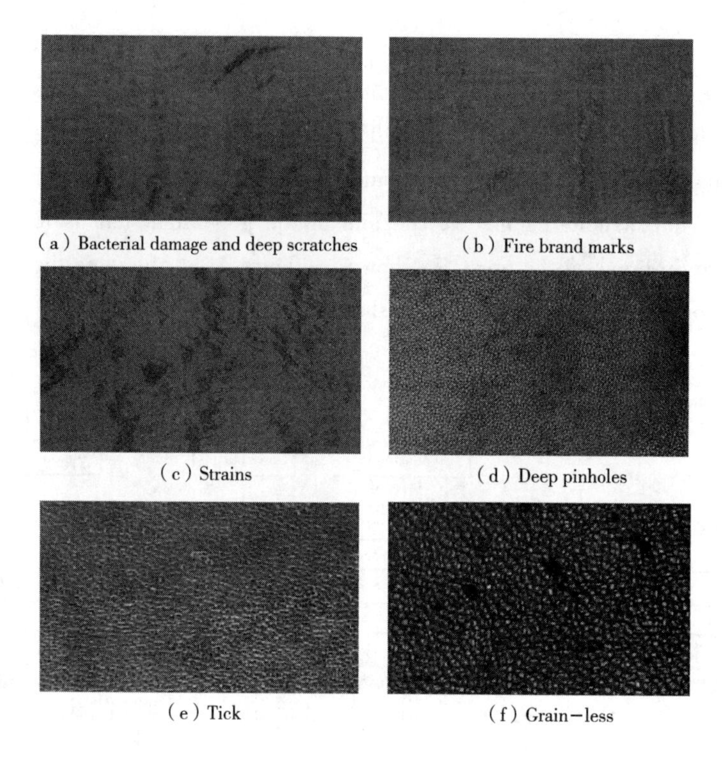

(a) Bacterial damage and deep scratches (b) Fire brand marks

(c) Strains (d) Deep pinholes

(e) Tick (f) Grain-less

Figure 3.53 Some kinds of defects on the surface of the crust

According to the state of the crust and the requirement of the finished products, the usages and the proper methods are chosen and applied in the process. The leather having covering-defect is adequately dried and buffered.

3.6.3.2 Spraying dyeing/stain coat

For aniline, semi-aniline type leather or suede, spraying dyeing/stain coat operation can be used to improve the levelness and brightness of the color of the whole crust, and further reduce the dyestuff liquid and resins so as to keep the natural hand of the leather. The formulation of the spraying dyeing includes dyeing liquid, organic solvent, penetrant and water. Dyeing liquid can give a different color for the surface of the crust or suede or increase the brightness of the color. Organic solvent, penetrant can help the levelness of the color and the penetration of the dyeing liquid in the crust. For instance, the formulation of the spraying dyeing is as follows:

Dyeing liquid	10-20
Penetrator	30
MOP	30
Water	250

3.6.3.3 Impregnation coats

In order to reduce the penetration of finishing slurry and maintain the softness of the crusts, improve the loose grain, reduce the "Orange Pell" effect and modify the tightness of the grain of the crust, impregnation coats are used for the finishing of the leather by the spraying operation which is sealer coat, filling, etc. It can be carried out by spraying or rolling operation.

The finishing slurry is made up of pigment, soft polymers, fillers, wax or other finishing auxiliaries and water. The pigment can give the relevant color to the crusts and have good covering properties. Soft polymers may be acrylate resins, polyurethane resins, or proteins, which can improve the break of the grain, the abrasion resistance of the leather and act as a sealer coat for porous leathers. As previously mentioned, fillers, wax or other finishing auxiliaries can give relevant properties for the impregnation coats.

The finishing slurry used for it may be the anionic or cationic. In general, cationic layers have more covering properties and better hand than those of the anionic layers. Impregnation coats will be dried and thenironed by an ironing machine under the high temperature so that the coat can further improve its abrasive and penetration resistant properties. If the water absorption of the crust is not so good or strong, impregnation coats can be deleted and directly finished with a base coat by the roller or paddle operation. If the water absorption of the crust is poor or bad, the impregnation coats should contain the abrasive resin or increase the abrasive layer so that it helps to form the strong abrasive layer between the crust and the finishing layers.

3.6.3.4 Base coats

Base coats are primarily concerned with coloring the leather, obliterating or partially hiding all the irregularities of the grain, and forming a foundationof the general finishing layers.

The base coat contains dye solution and pigments, film-forming binders, fillers, matting or dulling agents, wax or other auxiliaries. Dye solution and pigments can color the base coat. The dye solution can penetrate into the leather and provide a staining effect for the film. The required color and its good covering power can be obtained by matching with different pigments. Film-forming binders such as an acrylic binder, polyurethane resin, acrylic-polyurethane compound, protein film-former (modified casein), etc can be used in the base coat. But these film-forming binders must all be soft and well plasticized so that the finished film can conform with the distortion of the leather. These film-forming binders should have good or certain penetrating and adhesion properties so that the base coat is firmly anchored to the leather. The filler can give good leveling action and covering power for the base coat. Matting or dulling agents are frequently used in the base coat in order to cover the flaw of the crust by reducing the light transmission through the finished film and increasing the opacity of the finished film. In addition, amorphous waxes and water-soluble resins are both used for improving the covering properties. Filler, wax, modified casein, etc can give the leather plate release properties. When it is desired to increase the heat and solvent resistance of the base coat, it is generally practiced to add a little casein or polyamide materials to the finish.

After the application of the base coats, the leather may be given a little heat and pressure in order to consolidate the grain surface further.

3.6.3.5 Middle coats

The middle coats are generally a little harder than the base coats and contain a smaller quantity of coloring agents. Their function is to complete the desired degree of leveling of the color and to form films of the required gloss and abrasion resistance of all kinds. They are generally more dilute than the base coats, and the binders and seasons are modified to give the harder films required in this part of the finishing system. The middle coats are generally applied by spraying manually or automatically.

3.6.3.6 Seasoning coats

The seasoning coats provide a required gloss film by matching with different resins and the application of glazing, plating, and ironing at certain temperature and pressure. The film-forming materials employed are hard and non-tacky. The seasoning coats contain a minimum of or no coloring agents.

Glazed leathers are given a protein-based season, while most other types of finish are given seasons based on resins, celluloses, waxes, and mucilaginous materials. The season plays an important part in controlling the abrasion resistance, particularly the wet and dry rub-resistance of the finished leather. Most seasoning materials have to be capable of forming water-resistant films. The seasoning coat is generally applied by spraying and can be amalgamated into either the middle or top coat.

3.6.3.7 Top coat

Except for the same general functions ofseasoning coats, the top coat may have some special effect on the leather and produce a unique handle or appearance, or give some entirely new properties to the leather such as special pearlized, metallized, and patent effects. In addition, a number of water repellents, particularly silicones and fluorocarbons can be applied in top coat so as to obtain the water-proof or water-repellent effect of the finished leather (Figure 3.54).

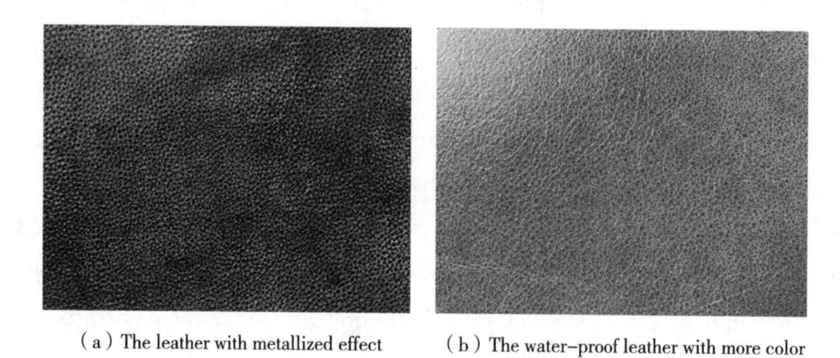

(a) The leather with metallized effect　　(b) The water-proof leather with more color

Figure 3.54 The leather with different styles

3.6.4 Operations of finishing of the leather

3-13

Machine operations are very important for the quality of the finishing of the leather so that it can be to determine the thickness, flatness, glossiness, etc of the finishing coats. In general, the methods of finishing operations include spraying, brushing and padding, roller coating, and curtain coating. Different finishing operations can different dosages of finishing materials and different properties.

3.6.4.1 Spraying operation

Spraying operations are applied with spray guns, either manually with spraying guns or automatically on a spraying machine. Spraying guns manually, the operating principle and the operation of spraying gun are shown in Figure 3. 55 and Figure 3. 56.

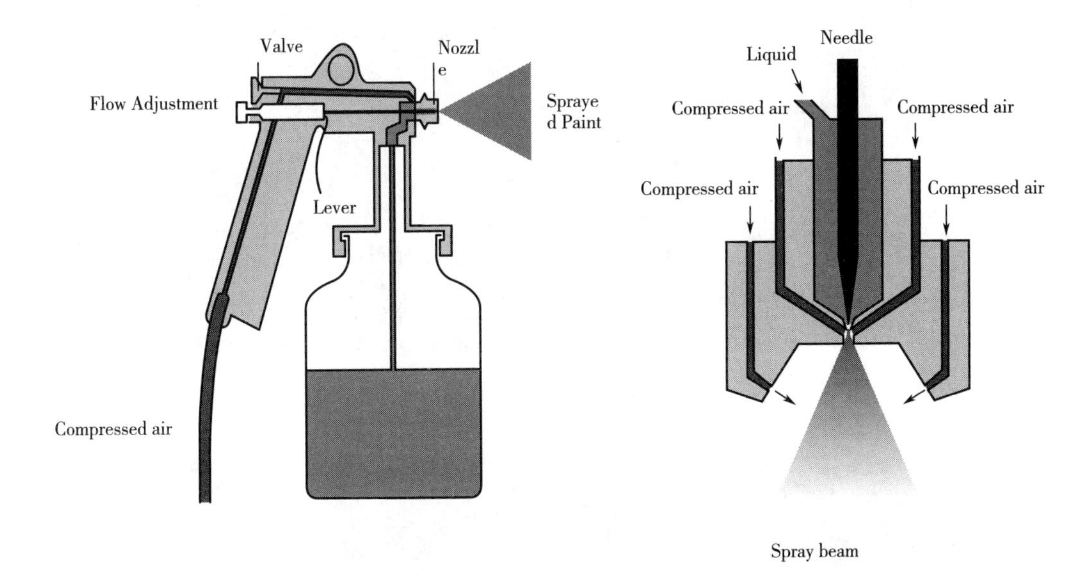

Figure 3. 55 Spraying gun manually and its operating principle

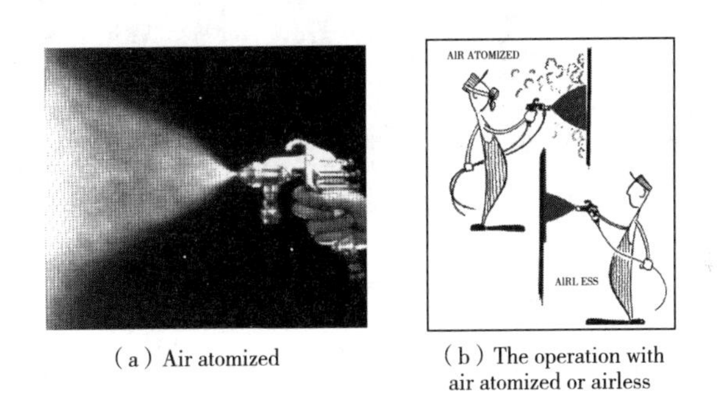

(a) Air atomized

(b) The operation with air atomized or airless

Figure 3. 56 The operation of spraying gun with air atomized or airless

The finishing solution can be sprayed on the surface of the leather by spraying guns with airless. But the finishing layer is thicker than that of air atomized.

Spraying guns in the spraying machine and the spraying machine are shown in Figure 3. 57 and Figure 3. 58.

In general, spraying guns manually will be used for the finishing of the small batch of leathers or finishing in the laboratory. Whereas the spraying guns automatically and their equipment will be applied for the finishing of big batch of leathers during the production of leather. Therefore in most cases spraying machines are used instead of hand spraying for bulk production.

In the processing of bulk production, the flat leather is fed on the nylon band conveyor travelling at the speed of 40–70ft. per minute and passes beneath the spraying guns. Then the finishing solution is fed to spraying guns by gravity or compressed air, and is ejected from a fine nozzle or orifice by

Figure 3.57 Spraying gun in the spraying machine

Figure 3.58 Spraying machines and spraying automatically

pressure, as a spray of fine "atomized" drops, which are blown onto the surface of the leather. The guns are automatically controlled so that they only spray the leather that is passing directly beneath them. In most cases, it is desirable that these drops should then flow together to give a continuous wet film on the leather before drying. Failure to flow out may result in poor anchorage, poor rub fastness, loss of gloss, poor film strength, or non-smooth film surfaces, e. g. "orange-peel" effects. The sprayed leather is dried quickly to a non-tacky state in a few minutes in the drying tunnel.

Spraying operation can be used in all the finishing coats including impregnation coats, base coats, middle coats, top coats, touch coats, etc.

3.6.4.2 Brushing and padding

The brush used for the finishing of the leather is like a soft-haired shoe brush and the pad is a wooden block covered with plush or a piece of sponge with gauze, as shown in Figure 3.59.

The finishing solution is painted on the surface of the leather by the brush or pad dipped into the paint. The application by brush and pad has a pronounced effect in speeding up the wetting and hence hydration of the surface fiber. The finishing solution should have good leveling properties so that it can

Figure 3.59 The brush and pad used for the finishing of the leather

provide uniform coats for the leather after it is dried. If coagulation is not very rapid, the padding ensures good uniform penetration of the binder into the surface and good adhesion or anchorage of the finish. If the coagulation is too rapid, the film will be deposited very superficially and may even be broken up or crumble with the action of the pad. Brushing and padding usually are easy to be controlled, and adjusted so that it is used for the finishing of base coats, middle coats, seasoning coats, etc. However, because the operation is inefficient, has bad uniformity and has big labor intensity so that it can be replaced by roller coating.

3.6.4.3 Roller coating

Roller coating is one of the most important finishing operations of the leather which is widely used during the finishing of base coats, topcoats, the coats with special effects of the leather by the roller machine. The roller machine contains the screen or pattern roll, rubber roll, feed doctor blade, doctor blade, convey or belt, etc, as shown in Figure 3.60 and Figure 3.61. The finishing solution can be applied on leather via rollers that have different types and corresponding dosages of the leather (Table 3.5). By using appropriate pattern rollers, it is possible at the same time to obtain special effects. Base coats and printed effects are applied with screen rollers or appropriate pattern rollers using the direct process. For roller coating without a pattern the indirect process is used. The finish is applied to the leather from a coating roller via a rubber roller and spreads over the leather, giving it a smooth surface.

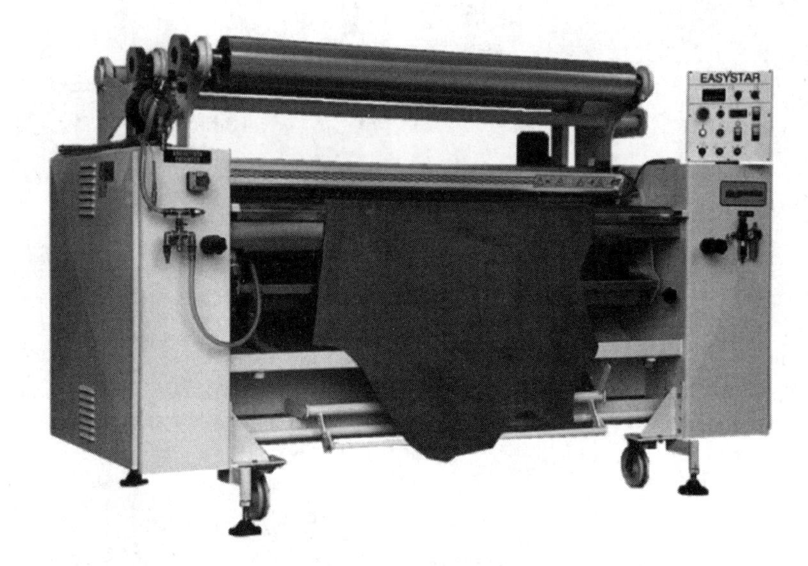

Figure 3.60 The roller machine used for the finishing of the leather

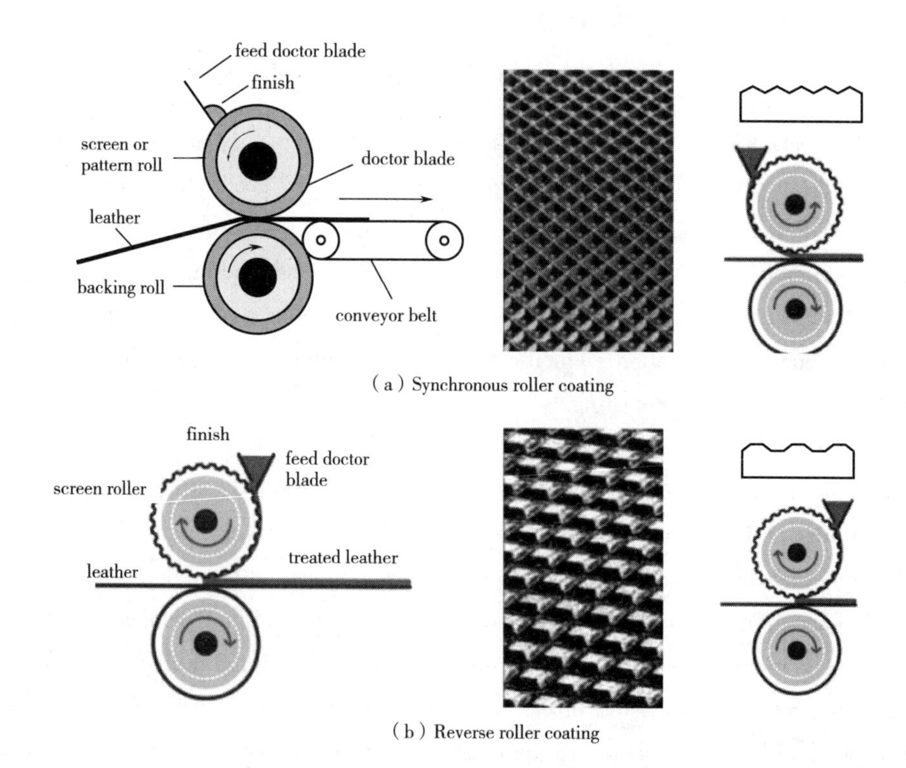

Figure 3.61 Two types of roller coating

Table 3.5 The finishing methods, type of rollers and their dosage of the roller machine

Methods	Type of rollers	Dosage of finishing/ (g/ft^2)	Methods	Type of rollers	Dosage of finishing/ (g/ft^2)	Type of rollers	Dosage of finishing/ (g/ft^2)
Synchro roller coater	6L	14-18	Reverse roller coater	10/B	24-33	40/F	1-2.5
	10L	10-14		10/C	18-27	14G	12-18
	16L	6-9		20/B	15-25	17G	10-16
	24L	4-7		20/C	12-18	21G	7-14
	40L	1-3		30/A	10-16	25G	5-10
	48L	1-2		30/X	8-12	30G	3-7
	60L	1		30/C	5-10	40G	2-5
	—	—		30/F	3-6	50G	1-2

It is worth noting that the finishing solution should have ideal viscosity in order to ensure the uniform coating. If the viscosity of the finishing solution is low, the leather painted by the roller machine has the "chicken feet" or "fish eye" print. The viscosity of the finishing solution can be determined by 4# Ford cup, as shown in Figure 3.62.

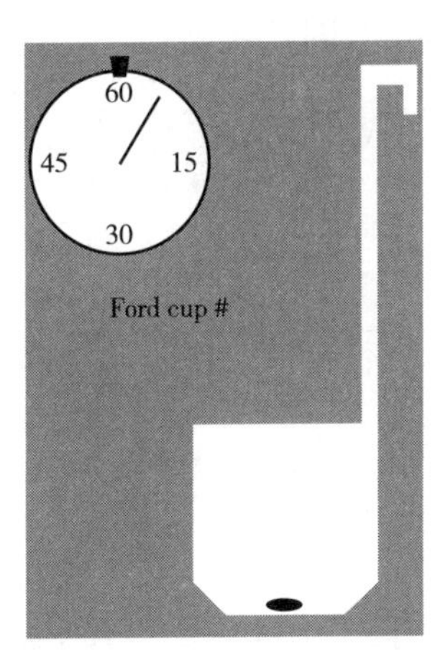

Figure 3.62 Ford cup used for the finishing of the leather

3.6.4.4 Curtain coating

Curtain coating is also one of the most important finishing operations of the leather which is used during the finishing of impregnation, the base coat of the leather by the curtain coater, as shown in Figure 3.63.

During the curtain coating, the finishing solution is pumped into a feed tank (head) with a curtain coating device positioned over a conveyor belt. The finish flows out of the tank in the form of a liquid curtain onto the surface of the leather, as it passes underneath and is taken up by the leather, as shown in Figure 3.64. Any finish, which does not land on leather, is collected and pumped process is

Figure 3.63 Curtain coater used for the finishing of the leather

particularly suitable if relatively large amounts of finish are to be applied, as it is economical, environmental in its product consumption and cost-effective.

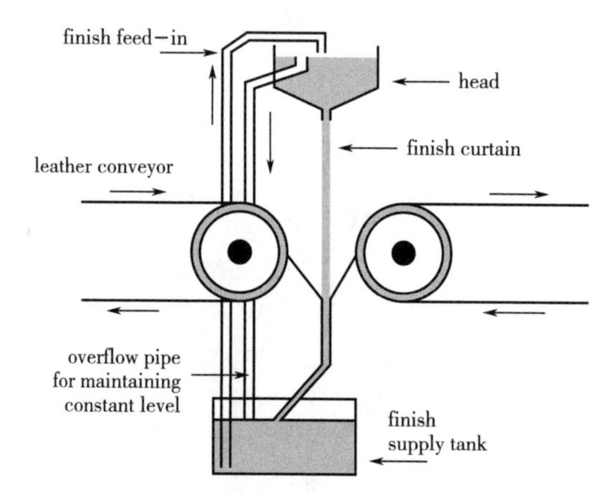

Figure 3.64 The operating principle of the curtain coater

3.6.5 Other operations in the finishing process

Except for the previous operations of finishing the leather, ironing/plating, pressing/embossing, grading, measuring operations, etc are also used in the finishing process.

3.6.5.1 Ironing/plating and pressing/embossing

As previously mentioned, the ironing/plating operation is the same as that in the softening process, it can improve the glossy, water-proof, abrasive properties and tensile strength of the leather. In addition, because the operation is carried out under certain temperature and pressure, it can speed up the forming of the continuous film of different finishing coats, and increase the adhesion between the crust and the impregnation coat or among different coats of the finishing. Thus leather with properties such as smooth, glossy, tidy, water-proof, etc is obtained. If the through feed rotary Ironing machine is replaced with a flat iron, the operation is plating. During the operation, the temperature, pressure, and conveying speed of roller time can be adjusted in order to meet the requirement.

In order to produce leather with varied grain texture or fashion style, the embossing operation is the one of important methods which is carried out on embossing machine with embossing roller or plate. Through feed rotary embossing machine is shown in Figure 3. 65. When the finished leather is conveyed to the heated roller or placed under the heated plate with certain pressure, the decorative pattern on the embossing roller or plate will also emerge on the surface of the leather. Likewise ironing/plating, the temperature, pressure, and conveying speed of roller time can be also adjusted in order to meet the requirement.

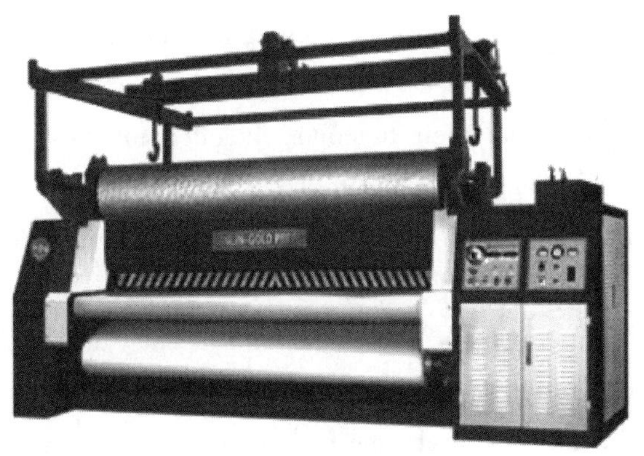

Figure 3. 65 Through feed rotary embossing machine

3. 6. 5. 2 Grading and measuring

The object of grading operation is to determine thequality of the finished leather. Leathers are graded for temper, uniformity of color and thickness, and the extent of any defects which appear on the leather's surface. The most perfect leather naturally commands the highest price and its ultimately cut into the finest leather articles.

The aim of measuring operation is to determine the area of each leather because the finished leather is sold on the basis of its area. Measuring operation of the leather can be achieved by the measuring machine, as shown in Figure 3. 66. When the leather is conveyed to the beam controlled by the computer, the computer can automatically record and print the area of the leather.

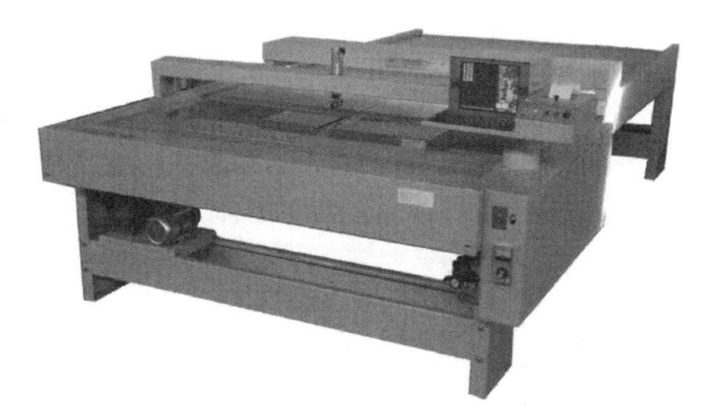

Figure 3. 66 A full-automatic measuring machine with computer

Questions

(1) Briefly describe the stages of operation and their functions in leather manufacture.

(2) Describe the processes and their functions included in the beam-house operation.

(3) Describe the aims of soaking process, adopting the chemicals and affecting factors.

(4) Describe the aims of degreasing process, adopting the chemicals and affecting factors.

(5) Describe the components of hair, conventional unhairing methods and their mechanism.

(6) Describe the aims of unhairing and liming process, adopting the chemicals and affecting factors.

(7) Describe the aims of deliming process, adopting the chemicals and affecting factors.

(8) Describe the aims of bating process, adopting the chemicals and affecting factors.

(9) Describe the processes and their functions included in the tanning process, adopting the tanning agents, the tanning effect and the characteristics of the relevant leather.

(10) Describe the aims or functions of pickling process and the relation between the pickling process and tanning process.

(11) Describe the characteristics of chrome tanning agents, the mechanism, effects of chrome tannage and the affective factors in the tanning operation.

(12) Describe the characteristics of chrome-free mineral agents, the tanning mechanism, effects and the characteristics of the relevant leather.

(13) Describe the structure, sorts and characteristics of vegetable tannins, the mechanism, effects of vegetable tannage and the affective factors in the tanning operation.

(14) Describe the structure, sorts of syntan, their mechanism, effects and the affective factors in the tanning operation.

(15) Describe the structure, sorts of aldehyde tanning agents, their mechanism, effects and the affective factors in the tanning operation.

(16) Describe the processes and their functions included in the dye-house operation.

(17) Briefly describe the functions of neutralizing process for the further processing of the chrome-tanned leather.

(18) Why is the retanning process considered as the golden touch in modern leather manufacture?

(19) What are the frequently-used retanning agents, their structures and the properties?

(20) Describe the physical structure and chemical characteristics of the leather and the affection for retanning, dyeing and fat-liquoring process.

(21) What are the aims and requirements of dyeing process?

(22) Briefly describe the difference between dyestuff and pigment.

(23) What is the frequently-used dyestuff, their structures and the properties?

(24) Briefly describe the dyeing process and the affecting factors.

(25) Describe the methods and function of fat-liquoring process.

(26) Describe the frequently used fat-liquoring agents and their characteristics.

(27) Briefly describe the mechanism of fat-liquoring process and affecting factors.

(28) Describe the aims or functions of the drying methods and the existing forms of water in the crust.

(29) Describe the drying methods for the leather and their characteristics and affecting factors.

(30) Describe the aims or functions of the finishing operation.

(31) Briefly describe the composition of finishing agent and related functions.

(32) What is the mechanism of the form-filming materials or finishing agents?

(33) Briefly describe the rule of the formulation of finishing process and design the finishing formulation of furniture leather, shoe upper leather or garment leather.

(34) Describe the mechanical operation and affecting factors in the finishing process.

Chapter 4 Clean technology of leather manufacture

As previously mentioned, leather industry is a traditional industry with comprehensive advantages, long history and good trend of development because it can achieve the resource utilization of animal hide or skin originated from husbandry, and meets people's needs for leather products. In recent years, leather industry has achieved rapid development in the world. China's output of leather accounts for about 30% of global output, and the output of double-face leather is about 75% - 80% of that of the global. Therefore the leather industry play an important role in the development of the national economy.

4-1

Leather industry includes leather manufacture (leather-making), double face leather-making, leather products, leather chemicals, leather machinery and leather hardware. Leather manufacture (leather-making), and double face leather-making are the base of leather industry and play an important supporting role in the healthy development of Leather industry. But with the development of the national economy and improvement of people's living standards, the sustainable development of leather industry faces severe challenges because of environmental pollution. These pollutants include chromium (Ⅲ or Ⅵ), sulfide, salt, organic matter, ammonia nitrogen (NH_3-N), volatile organic compounds (VOCs) and solid waste, etc which originate from the curing of hide or skin, beam-house, tanning process, retanning and dyeing, finishing, etc. In addition, the composition and content of these pollutants are relevant to the kinds of hide or skins, the technology of leather-making and the sorts of leather products. Hence based on requirements of the policy of national industry, environmental protection (ecological environment), people's requirements for a better life and improvement of competitiveness in international markets, research and implementation technology of clean production of leather manufacture has a very important practical significance.

The aim of clean production of leather manufacture is to reduce or get rid of the pollution during the processing of leather making. In addition, the reutilization of the effluent and waste solids should be reused by the new technology. In the chapter, the technology of clean production of leather manufacture is introduced. However, the technology of reutilization of the effluent and waste solid will be discussed in the next chapter.

4.1 Clean preservation technology of hide or skin

The quality of hide or skin directly affects the one of the finished leather. Hide or skin contains protein, greasy, water, etc which provide abundant nutrition for the growth and reproduction of the microorganisms. The protease produced by the microorganisms decays the hide or skin. Therefore the preservation and its quality of hide or skin are very important for the utilization of raw materials and the improvement of the finished leather.

The basic principle of the preservation of hide or skin is to control the content of water, temperature, pH, adding antiseptic, etc which can form the environment which is not suitable for the growth and reproduction of bacteria. As previously mentioned, the traditional methods include drying, salt curing, chilling, etc. Different methods have different some advantages and other disadvantages. The drying process can reduce the content of water in the hide or skin to 14%-18% and restrain the growth and reproduction of bacteria. The operation of drying process is simple and low cost, but collage fibers is easy to be damaged and the quality of the finished leather is poor. The chilling process can ensure the quality of hide or skin, but the chilling process has high energy consumption and shipping costs, etc. Salt curing possesses the advantage of being easily operated, low price and wide range of applications which acts as one of the main preservation of hide or skin. But massive salt is applied in salt curing so as to cause the pollution of salt and high the content of total dissolved solids (TDS) in the effluent.

In recent years, governments make the increasingly stringent standards for effluent discharge which include the standards of the effluent with salt or the content of chloride ion (Cl^-). Therefore in order to reduce or get rid of the pollution of salt, the research and application of clean preservation of hide or skin have significance for clean production of leather manufacture. Researchers have studied many technology of the preservation of hide or skin whose aims ensure the quality of hide or skin and environmentally friendly. These clean preservation of hide or skin include curing with less salt, potassium chloride (KCl) or silicate, chilling, radiation, drying, etc.

4.1.1 Curing with less salt

Traditional salt curing has some advantages such as low price, good curing effects, etc. But the curing causes the pollution of salt. Nowadays salt curing is difficult to be completely cancelled. The curing with less salt is studied by some researchers. The curing using salt with bactericide, bacteriostat, dehydrant or other reagents will reduce the dosage of salt, salt pollution and preserve the hide or skin.

4-2

One of the research on curing with less salt is the method combining salt and boric acid. Boric acid can absorb the water among the gap of the fibre and has sterilizing effects. Hughes I R studied the methods of preservation of hides by combining saturated boric acid solution (4.5%) and saturated sodium chloride solution. Results show that hide can be preserved for a week or so under the condition of 30℃. Kanagaraj J, Sundar V J, Muralidharan C, et al did a series of anti-corrosion experiments by smearing the flesh side of hide with the mixture of boric acid and sodium chloride under the condition of 30℃. Results show that the optimum ratio is 2% boric acid and 5% sodium chloride. The results of poilt production in Tannery show that the value of BOD, COD, TDS, TSS and Cl^- of soaking effluent of hide preserved used by the mixture of 2% boric acid and 5% sodium chloride for two weeks are all below that of traditional salt curing and the quality of the finished leather is equivalent to that of traditional methods, as shown in Table 4.1 and Table 4.2. The price of preservation of hides is slightly higher than that of traditional salt curing and the methods can solve part of the problem of salt pollution. But the effluent containing boric acid will damage soil structure so that boric acid can't be used in the preservation of hide or skin. The mechanism of boric acid in the preservation of hide or skin can have an inspiring significance for studying the substitute for salt. In addition Russell, et al studied a way of curing with less salt which smears the flesh side of hide with a mixture of bactericides. The bactericide consists of EDTA-Na (25%), sodium chloride (40%) and sawdust (35%) whose dosage is about 150g for a piece of sheepskin or 3,000g for a piece of bovine hide. The weight of the hide or skin cured by the bactericide is lighter than that of traditional salt curing. Results of pilot production show that LIRICURE technology is feasible by a comprehensive consideration of the price of production, sewerage and environmental benefits. The curing is suitable for the short-term preservation of hide or skin because the storage period of the hide or skin cured by LIRICURE technology is four weeks. In addition, Ca^{2+} and Mg^{2+} in the water can weaken the bacteriostatic effect of EDTA-Na.

Table 4.1 The value of the relevant item of soaking effluent

Item of testing	Boric acid-salt curing	Traditional salt curing
BOD/(mg/L)	9±1	9.5±1
COD/(mg/L)	17±1	26±1
TDS/(mg/L)	45±2	264±5
TSS/(mg/L)	10±0.5	21±1
Cl^-/(mg/L)	22±1	195±5

Table 4.2 The mechanical properties of the leather originated from different curing

Item of testing	Boric acid-salt curing	Traditional salt curing
Tensile strength/($N \cdot mm^{-2}$)	20.6±0.5	20.1±0.5
Elongation at break/%	45±3	48±2
Tear strength/($N \cdot mm^{-1}$)	29±2	27±2
Burst strength/N	196±5	216±5
Thickness/mm	10±0.5	10±0.5

4.1.2 Curing with potassium chloride

Sodium chloride is one of the most difficult components removed in the effluent which can cause soil salinization and crops cannot grow if the effluent is directly discharged. Potassium chloride is a fertilizer for supporting the growing of plants. If potassium chloride can replace sodium chloride in the preservation of hide or skin, potassium ions in the effluent are absorbed by crops and promote crop growth. Bailey did plenty of research and thinks the curing technology of potassium chloride taking the place of sodium chloride is achieved. He treated hide or skin by adopting 3mol/L, 4mol/L, and 5mol/L potassium chloride solutions respectively under the condition of different temperatures. Results showed that the quality of the hide or skin cured by 4mol/L, 5mol/L potassium chloride solution is very good. The storage life of the hide or skin cured by potassium chloride can obtain more than three months. Some suggestions are that a little bacteriostatic agent is added during curing hide or skin with potassium chloride.

Nowadays, there are a series of problems with potassium chloride replacing sodium chloride because of the price, solubility, etc.

4.1.3 Curing with silicate

During the methods of the preservation of hide or skin, silicate can replace sodium chloride in many countries. Results show that the hide or skin cured by silicate has excellent storage and the content of TDS, salt of the soaking effluent are significantly decreased/reduced. The methods have been applied on the semi-industrial scale. Silicate can be used in the curing of hide or skin by drumming or powdering. Drumming is that hide or skin is soaked in 5%-30% sodium silicate solution with 100%-150% water for 2-5 hours. Then drain and neutralize to pH 5.0-5.5 with acid, at last horse up. The cured hide or skin can be stored for many months. Powdering is when the white powder is smeared on the flesh side of hide or skin whose dosage is 20%-25% of the weight of raw skin. The white powder is obtained by sodium silicate solution and is neutralized with formic acid and sulfuric acid, then washing and drying.

The hide or skin cured by silicate is very hard and contains 10%-15% moisture which is easy to be soaked. The quality of the finished leather is equal to that of the traditional salt curing.

The price of silicate powder is twice that of sodium chloride. With comprehensive consideration of the price of silicate powder, water treatment, shipping, etc, the curing methods of hide or skin are feasible. It is noted that the particle size of silicate should be ideal. The particle size of silicate is too fine and is easy to be packed so that it weakens the antiseptic effect. The particle size of silicate is too big so that it affects the distribution of silicate.

Besides curing with less salt, potassium chloride (KCl) or silicate, chilling, radiation, drying, etc is adopted during the storage of hide or skin.

4.2 Clean technology of beam-house operation

Beam-house operation is one of the most important procedures during the process of leather manufacture and the base of leather-making. Beam-house operations include sorting, soaking process, degreasing process, unhairing process, liming process, reliming process, deliming process, bating process. Beam-house operation is the main source of pollutants and wastes in leather industry. Many data show that BOD, neutral salt, the concentration of suspended solid (SS) produced in beamhouse accounts for over 80% of the whole process. In order to solve the pollution fundamentally, we should adopt clean production of beamhouse operation, especially the unhairing process and deliming process so that it can reduce the production and discharge of pollutants. Some novel technologies of clean production are developed to decrease pollution to a certain extent.

4-3

4.2.1 Technologies of decreasing pollution in soaking process

The aim of soaking process is to cause the cured hide or skin to revert to the state of fresh hide or skin as much as possible and lays a good foundation for subsequent processing. During the soaking process, some soaking auxiliaries are added into the process in order to shorten the time, increase the even of soaking, remove the non-collagen components and protect the collagen. Meanwhile, the soaking auxiliaries such as acid, alkali, salt, bactericide and others also cause pollution of the environment. Therefore the development of some soaking auxiliaries with less or no pollution has become the research interests of clean technology of soaking process.

Nowadays, acting as one of the most important soaking auxiliaries, enzymes can speed up the wetting of hide or skin fibers, partly hydrolyzing the fat, interfibrillar substance, hair follicle, etc, opening up the ruffle of hide or skin and increasing the gain of leather. Therefore enzymes can take the place of some heavily polluted soaking auxiliaries wholly or partly in the soaking process. The main ingredient of the enzyme used in the soaking process is mainly alkali protease which can remove the non-collagen well. For instance, Novolase SG (Novozymes), Erhazym C (TFL), Aglntan PR (Schill & Seilacher), Trupy MS (TRUMPLER), DOWELLZYM S2 (DOWELL), etc has good effect in soaking. In addition, proteinase and lipase can be simultaneously used in the soaking process because they can form the synergistic effect and promote the soaking.

It is noted that we should understand the main ingredient of the enzyme and its parameters during the application. Then based on the state of hide or skin, characteristics of processing, needs of the finished products and the cost of production, the enzyme will be chosen and used in optimum conditions such as consumption, pH, temperature, etc.

"Proviera Biotech" proposes 100% natural and biodegradable soaking agents from probiotic technology that significantly reduce the carbon footprint and the water usage during the manufacturing compared with traditional synthetized surfactants derived from the crude oil industry. Additionally, they can reduce the COD of the soaking effluents. The product range consists of ProSoak (soaking and wetting back agent), ProSpread (dispersing agent) and ProDegreaze (degreasing agent). The are highly efficient and help tanner to reduce costs, improve leather quality while reducing the environmental impact.

4.2.2 Technologies of decreasing pollution in degreasing process

The grease in the hide or skin affects the penetration of water-solubility chemicals so that the chemical reaction between water-solubility chemicals and hide or skin collagen is reduced in the process of tanning, retanning, dyeing, etc. Therefore the degreasing process is very important for the whole production of leather manufacture and the quality of the finished leather, especially sheepskin, pigskin and hide or skin with more grease.

During the traditional degreasing process, saponification, emulsification and solvent methods are adopted. Saponification can effectively remove the grease on fur and the surface of hide or skin, yet the results of the degreasing are less for the inner grease of the hide or skin. However, emulsification and solvent methods have good degreasing properties for the grease distributed in whole hide or skin.

Nowadays, saponification and emulsification methods are widely applied in the leather-making process. Different kinds of surfactant are used in the emulsification methods. Some surfactants consisting of the structure of alkylphenols are difficult to be degraded so as to pollute the environment. Therefore the surfactant with being easily degraded should be chosen and applied in the degreasing process in order to protect the environment and achieve sustainable development. ProDegreaze and ProDegreaze Plus, from probiotic fermentation, can also replace degreasing agents and have a good effect striping off natural fat. In addition, the grease in the hide or skin can be hydrolyzed by enzyme and be removed. Compared with acid, alkali, and surfactants, enzyme is environmentally friendly chemicals. Alkaline lipase such as Erhazym LP (TFL), Greasex series (Novayme), DOWELLZYM HK (DOWELL), etc are often used in the beamhouse operation for degreasing which has good activity and stability under the condition of alkali and 35℃. ProDegreaze can be combined with lipases obtaining a synergistic effect and improving the degreasing efficiency.

Moreover, ultrasounds and supercritical carbon dioxide are also used in the degreasing process. Ultrasounds can destroy the fatty cell, emulsify and disperse degrease and improve the degreasing effect. Sun danhong, et al studied the effects of ultrasound of 20 kHz and 40 kHz on degreasing of pigskins and goatskin. Results showed that ultrasound can promote the dissolution of grease in raw skins and the removal of epiderm, favors removal of soluble proteins and grease in raw skins. While the structure of collagen is well-kept in this process. Sivakumar, et al studied the degreasing process with solvent under the ultrasound conditions. Results showed that ultrasound can assist the emulsifying the grease and improve the effect of degreasing so that it can eliminate the pollution of solvent, and reduce the consumption of surfactant. Marsal et al studied the degreasing of sheepskin with supercritical carbon dioxide. Results showed that the degreasing effect was related to the concentration, and velocity of flows of carbon dioxide, reactive time and the content of water in the hide or skin. The degreasing effect arises along with the enlarging of concentration, and velocity of flows of carbon dioxide. The degreasing effect can achieve over 94% under the optimum condition.

The technologies of degreasing with ultrasounds and supercritical carbon dioxide are still studied on the scale of the experiment. We believe that the technologies will be applied in leather-making in the future along with the development of research.

4.2.3 Technologies of decreasing pollution in unhairing and liming process

The aims of the unhairing and liming process are to remove the fur on the hide or skin and loose the collagen fibers which directly affect the softness, fullness, elasticity, state of grain, physical and mechanical properties, and gain of the finished leather. As previously mentioned, sodium sulfide, sodium hydrosulfide, lime and other auxiliaries are applied to the traditional unhairing and liming process. Therefore traditional technology of unhairing and liming has the characteristics of low price, being easy to operate and control, stability of the quality of the finished leather and being applied widely, etc so that the technologies are applied in many tanneries in the world. But the unhairing and liming process brings about the pollution of sulfide, organic compounds, lime and other pollutants for the environment. How to reduce pollution is always an important topic for researchers in the field of leather manufacture.

Hair-preserving technologies are one of the most important methods for the clean technology of the unhairing process. There are two hair-preserving technologies including liming directly and sweating until 1880. The technologies of unhairing based on sodium sulfide and lime were widely used in the many tanneries in the world from 1880 to now which save time, labor, open up the collagen fibers and clean the pelt. But the technologies bring about the pollution including effluent, sludge and others. Therefore in order to reduce the pollution from sodium sulfide, hair-preserving technologies have been further studied and gradually are applied in the production of leather in recent years. These technologies include CSIRO, HS, BLAIR, unhairing assisting with enzyme and enzymatic unhairing. The process of CSIRO, HS and BLAIR unhairing mainly base on the sodium hydrosulfide, sodium hydrosulfide and lime.

During the process of CSIRO unhairing, the hide or skin is firstly dipped in the solution of sodium hydrosulfide in order to ensure the hair-preserving properties of HS^-. Then sodium hydrosulfide adhering to the surface of hair is removed by calcium hypochlorite so that it can remain the strength of the hair. When the pH of the solution is above 12 and under the condition of the presence of lime, the strong reduction of sodium hydrosulfide in the hair root can weaken or destroy the connection between hair and hair follicle and be removed under the condition of mechanical action. Then the hair is separated from the effluent by being filtered. The effluent can be reused in the liming process.

The technology of CSIRO unhairing process is shown in Table 4.3 which is feasible and can be applied in the production according to the state of pelt, hair and the recovery rate of hair (>90%).

Table 4.3 The technology of CSIRO unhairing process of bovine hide

Process	Chemicals	Offer/%	T/℃	t/min	pH	Remarks
Soaking and degreasing process						
Dipping	Water	30	18-22			
	Sodium hydrosulfide	0.7		120		Drain
Washing	Water	40	18-22	5		
Hair-preserving	Water	30	18-22			
	Calcium Hypochlorite	0.1		6		
Unhairing	Lime	1		30+50	12.0-12.5	
Reliming	Water	50	18-22			
	Sodium sulfide (60%)	2	18~22	15		
	Lime	1		360		
	Water	100		60-120		Running 5 min every 60 min
Drain, filter, washing, splitting						

The HS unhairing and liming system is the technology of hair-preserving with less sulfide, which was developed by TFL Co. Ltd and can be applied in the drum and paddle. The technology of HS unhairing process is shown in Table 4. 4. Erhavit HS has very little effect on the hair shaft at the range of 9. 5 and 10. 5. However, Erhavit HS can only act on the hair root when the pH of the solution is 12-13 and under the condition of the presence of lime. Then sodium hydrosulfide is added to destroy the hair root and the hair is removed. The HS unhairing and liming system can increase the gain of leather, clean the surface of the pelt and be suitable for the production of leather manufacture.

Table 4. 4 The technology of HS unhairing process of Bovine hide

Process	Chemicals	Offer/%	T/℃	t/min	pH	Remarks
Washing	Water	120	27	60		Drain
Soaking	Water	120	27			
	Erhazym S	0. 2				
	Borron ANV	0. 2				
	NaOH (50%)	0. 5		240	9. 5-10. 5	Drain
Unhairing	Water	80	18-22			
	Erhavit HS	1. 0		30		Running 30min, stopping 30min
	Lime	1				
	Sodium hydrosulfide	0. 7		75-90		Filtering
Liming	Water	50	27			
Washing	Water	40	18-22	5		
	Lime	2				
	NaOH (50%)	0. 5		30		Running 1 min every 60min, total 12 hours
Drain, filter, washing, splitting						

During BLAIR unhairing process, the hair shaft is treated so as to cause the hair-preserving properties. However, the hair roots aren't protected and are destroyed after sodium hydrosulfide is added into the unhairing system. Then the hair is removed by mechanical action. The technology of the unhairing process is shown in Table 4. 5. The hair-preserving action of lime is palliative and easy to be controlled. The lime can not only protect the hair shaft, but also slightly swell the hide or skin which can open up the hair pores and is in favor of the penetration of the subsequent chemicals. The pelt is clean and has even swelling. The technologies obtain 90% of whole hair, and reduce the content of protein in the effluent. The properties of the finished leather are equal to that of traditional unhairing technology. But it is noted that the control of the temperature and time in the technology is very strict because of being easy to occur the phenomenon of hair protection excessively. Therefore the implementation of the technology demands a high level of management in the tanneries.

Table 4. 5 The technology of BLAIR unhairing process of Bovine hide

Process	Chemicals	Offer/%	T/℃	t/min	pH	Remarks
Washing	Water	120	27	60		Drain
Soaking	Water	120	27			
	Erhazym S	0. 2				
	Borron ANV	0. 2		240		
Protecting hair	Lime	2		5+25		Running 5min, stopping 25min, twice
Unhairing	Sodium hydrosulfide	1		10+20		Running 10min, stopping 20min, twice
Washing & unhairing	Water	120	28-29	30-45		
Reliming	Lime	1				
	Sodium hydrosulfide	0. 5				
	FR-62	1		10+8×60		Running 5min every 60min, total 8 hours
Drain, filter, washing, splitting						

The technology of unhairing assisting with enzyme acted as a unhairing process with less sulfide which has been applied in the production of leather manufacture. The common technology of unhairing assisting with enzyme is a unhairing process with alkali protease and sulfide which can obtain a good unhairing effect by the synergism between alkali protease and sulfide. Compared with traditional unhairing process, the consumption of sulfide was decreased to 50%. The technology of the unhairing process is shown in Table 4. 6. Auxiliaries with enzyme, and surfactant added into the main soaking process can partly remove the grease and destroy the epidermis of hide or skin. Then liming auxiliaries added in the system can not only promote unhairing, but also assist to control the swelling, remove the grease and disperse the lime. The subsequent sodium hydrosulfide, lime, sodium hydroxide or thiourea peroxide can obtain good unhairing properties. If a little thiourea peroxide is used in the system, the consumption of sulfide will be decreased to 70%.

Table 4. 6 The technology of unhairing assisting with enzyme of bovine hide

Process	Chemicals	Offer/%	T/℃	t/min	pH	Remarks
Presoaking	Water	200	23	60		Drain
Main-soaking	Water	150	23			
	Auxiliaries with enzyme	X		120		

Continued Table

Process	Chemicals	Offer/%	T/℃	t/min	pH	Remarks
Liming	Water	70-80				
	Auxiliaries	1		20		
	Lime	1				
	Surfactant	0. 1		30-40		
	Sodium hydrosulfide/ Thiourea peroxide	2. 2-2. 5/ 0. 3-0. 4		40		
	NaOH (50%)	0. 5		40		
	Lime	2				
	Hair remover	0. 1		20+12×60		Running 5min, stopping 25min
Drain, washing, defleshing						

Some researchers in Sichuan University and Sichuan Dowell Co. Ltd developed a series of materials for unhairing, dispersants for collagen fibers and found SLF (sulfide-lime-free) hair-saving unhairing system. The technology of the SLF un-hairing process is shown in Table 4. 7. The SLF system is based on three unhairing agents (Dowellon UHE, Dowellon UHA, Dowellon UHB) and a dispersant for collagen fibers (Dowellon OPF). Dowellon UHE is an unhairing agent containing complex enzymes and has hydrolysis for theelastic fiber, grease, proteoglycan, etc distributed around the hair follicle. Dowellon UHA is an unhairing agent based on organic sulfur and has the properties of unhairing and adjusting swelling for the soaked hide or skin. Dowellon UHB has properties of unhairing and dispersing fibers for the soaked hide or skin. The SLF technology of unhairing can reduce the consumption of sulfide and lime so that the rate of hair removal is high and the pelt is clean.

Table 4. 7 The technology of SLF unhairing process of bovine hide

Process	Chemicals	Offer/%	T/℃	t/min	pH	Remarks
Washing	Water	150	23	60		Drain
Presoaking	Convention					
Main soaking	Water	150	23			
	Nowolase SG	0. 2				
	Nowolase DG	0. 2				
	Wetting agent	0. 3		60		
	NaOH (50%)	0. 5		30	9. 5-10. 5	Run 5min, stopping 25min, total 4-6h, drain

Continued Table

Process	Chemicals	Offer/%	T/℃	t/min	pH	Remarks
Unhairing	Water	80	23			
	Dowellon UHE	0. 3				
	Dowellon UHA	0. 8		30		Run 30min, stopping 30min
	Lime	1				
	Dowellon OPF	0. 3		60		
	Sodium hydrosulfide	0. 5				
	Dowellon UHB	0. 6		60		
	NaOH(50%)	0. 5		30		
	NaOH(50%)	0. 5		30		Drain, filter
Liming	Water	50	23			
	Dowellon OPF	0. 5				
	Dowellon UHB	0. 4		30		
	NaOH(50%)	1		30		Run 1 min every 60min, total 12 hours, drain

Enzymatic unhairing has a long history in leather-making. Compared with the traditional unhairing process, enzymatic unhairing can reduce 70% the pollution of sulfide and the content of COD, BOD in the effluent. But the technology is difficult to be controlled because the composition of enzyme is complicated. In recent years, as the need of clean production of leather manufacture and the development of the engineering technology of biological enzymes, the technology of enzymatic unhairing is studied again and is applied in the unhairing process. The mechanism of enzymatic unhairing is that the enzyme (non-collagenase) hydrolyzes the mucoprotein and endomucin among hair follicle, hair bulb, hair papilla and weakens the connection between the hair and the hide or skin. Then the hair is broken away from the hide or skin so as to complete the unhairing. The common enzymes used in the production of leather manufacture include 166, 1398, 3942, 209, 2709 in China. As the development of biological enzyme, keratinase is being separated and purified and being applied in the experiment.

The implementation methods of the technology of enzymatic unhairing are enzymatic unhairing with rolling and stacking, enzymatic unhairing with a certain temperature and water in the drum.

During the methods of enzymatic unhairing with rolling and stacking, a certain amount of enzyme, preservative and penetrant are added into the drum with the soaked hide or skin. Then the hide or skins are piled in pieces under the condition of the temperature until the hair can be pushed down. The effect of unhairing is affected by water, salt (ammonium sulfate), temperature, pH, other auxiliaries and pretreatment. The process of the technology is simple and easy to control which can obtain good quality of the finished leather. But the process coves large area, high labor intensity and long time. The technology is mainly used for the unhairing of pigskin.

During the methods of enzymatic unhairing with a certain temperature and water in the drum, a certain amount of enzyme is added into the drum and the hair is removed under the condition of optimum temperature, pH and concentration, etc. Because of the mechanical action and higher temperature, enzymes can rapidly penetrate into the pelt and hair, epidermis can be removed from the hide or skin in a short time. Because the effect of unhairing is affected by the state of hide or skin, water, temperature, pH, the concentration of enzyme, other auxiliaries, etc, the technology is difficult to control. The loose grain of the leather will emerge if the control of the technology is poor.

4.2.4 Clean technologies of dispersing collagen fibers

The degree of dispersion of the collagen fibers directly affects the softness, fullness, elasticity, the state of the grain, physical and mechanical properties of the finished leather. During the process of leather manufacture, the dispersion of the collagen fibers mainly depends on the liming process. Sulfide and lime are used in the traditional liming process so that the pH of the liquid stabilizes at 12.5 or so and the collagen fibers are fully dispersed. The technology has good advantages of low price, being easy to control. But because the solubility of lime is low and the use of sulfide in the liming process, it causes the pollution of the environment. Therefore clean technologies of dispersing collagen fibers are necessary for reducing or getting rid of the pollution of sulfide and lime.

In recent years, some new materials for decreasing the amount of lime have been studied and applied widely in the process of leather-making in order to increase the solubility, stability of lime or disperse the collagen fibers without lime.

In the materials for increasing the solubility and stability of lime, Mollescal PA, Feliderm K, etc have good solubility for lime by forming the soluble salt with calcium ions. In addition, NUE 0.6 MPX can be widely applied in most tanneries because it can increase the liming effect, improve the quality of the finished leather, increase the gain of the leather and reduce the amount of sulfide and lime. At the same time, NUE 0.6 MPX can help the liming materials penetrate and disperse in the hide or skin. Thus it is good for the protection of environment.

For the dispersion of the collagen fibers without lime, when different amounts of sodium carbonate, and sodium bicarbonate, and sodium hydroxide are separately used for the dispersion of the collagen fibers, results show that sodium hydroxide has the best dispersion effect. The optimized process is 1% sodium hydroxide (based on the naked hide or skin), 350% water in the drum and runs 1min every hour for six times and stops drum overnight. The degree of swelling of the pelt in the process is similar to that of the pelt used with 10% lime. Amylase is also used in the dispersion of collagen fibers because it possesses a strong selectivity for proteoglycan. Acted naked hide or skins as materials, when the amount of the amylase is 1.5% (based on the weight of naked hide or skin), water is 100%, and time of running is 180min, the degree of the dispersion of collagen fibers is equal to that of conventional technology. In addition, soluble silicate is studied in the the dispersion of collagen fibers. 1% sodium silicate solution can achieve pH 13 or so which is slightly higher than the pH of ideal swelling of hide or skin. Therefore silicate can be used for the opening up of the collagen fibril bundle. 1%, 2%, 3%, 4%, 5% sodium silicate (based on the weight of naked hide or skin) was used for the dispersion of collagen fibers in the drum and runs 5 minutes every hour (total of one day), stop drum overnight. The results of the applied experiment are followed in Table 4. 8. It can be seen from Table 4. 8 that the degree of swelling of the pelt with 1% sodium silicate much less than that of pelt treated by sodium sulfide and lime. However, the degree of swelling of the pelt with the range of 3% and 5% sodium silicate is excessive. The degree of swelling of the pelt with 1% sodium silicate is equivalent to that of conventional swelling. Therefore, the optimum amount of sodium silicate is 2% during the dispersion of collagen fibers. The chromium contents distributed in different layers of the wet blue treated by sodium silicate are similar and slightly higher than that of the conventional technology, which may be silicate has good dispersing properties for collagen fibers of the hide or skin and increase the absorptivity of Chromium. In the system of swelling with sodium silicate, COD, TS content of the effluent decrease 55%, 44% respectively, and no-sludge is produced. The cost of materials and labor is equivalent to that of conventional swelling. Therefore the technology of swelling with silicate is feasible in the process of leather-making.

Table 4. 8 Degree of dispersion of collagen fibers and the removal amount of proteoglycan

Methods of treatment	Increasing rate of the weight of the naked hide/%	Removing amount of proteoglycan/ (g/kg)
Lime 4%	24. 6	3. 27±0. 02
Sodium silicate 1%	18. 2	2. 40±0. 02
Sodium silicate 2%	25. 4	3. 35±0. 02
Sodium silicate 3%	28. 4	3. 42±0. 02
Sodium silicate 4%	30. 2	3. 38±0. 02
Sodium silicate 5%	30. 8	3. 44±0. 02

4. 2. 5 Technologies of pollution in deliming and process

Deliming is an important process along with the liming process whose aim is to remove the lime and alkali in the pelt, eliminate the state of swelling, adjust pH of the pelt and to create conditions for bating and pickling processes. In the deliming process, chemical methods are the main and washing is the auxiliary. In the process of leather-making, Acidic salts or weak acid organics are adopted in the chemical deliming process in order to ensure the process is in progress slowly and evenly. Acidic salts include ammonium sulfate and ammonium chloride, which is widely adopted because it has the characteristics of moderation of action, being easily controlled and low cost, etc. But ammonium salts produce large amounts of provocative and unhealthful ammonia which greatly improves the content of ammonia nitrogen (NH_3-N) in the effluent and causes air and water pollution. Therefore in order to reduce or abate the pollution of the effluent from the deliming process, cleaner deliming agents and their technologies have been studied and are gradually applied in the production. They are deliming technologies with carbon dioxide, magnesium salt, boric acid and less ammonia. The deliming technology with carbon dioxide can completely or partly take the place of ammonium salts abroad since the 1980s. The deliming technology with carbon dioxide can greatly reduce the nitrogenous compounds, stimulation from ammonia nitrogen and more than 50% BOD, whose mechanism is as follows.

$$Ca(OH)_2+CO_2 \longrightarrow CaCO_3+H_2O$$

$$CaCO_3+H_2O+CO_2 \longrightarrow Ca(HCO_3)_2$$

The reaction between lime and carbon dioxide forms insoluble calcium carbonate when the pH of the solution is 8.3. Whereas calcium carbonate can react with water and carbon dioxide and obtain soluble calcium bicarbonate so as to accomplish the aim of deliming. As shown in Table 4.9, compared with conventional the deliming technology, the deliming technology with carbon dioxide reduces the content of ammonia nitrogen (NH_3-N) in the effluent, clean and fine the grain of the pelt. The effect of deliming with carbon dioxide is related to the thickness of the pelt, the amount of carbon dioxide, pH of the solution, temperature, liquid ratio, etc. On the whole, the technology of deliming can efficiently diminish the content of ammonia nitrogen (NH_3-N) in the effluent and BOD, the air pollution and be easy to be controlled. The quality of the finished leather is good so that it can be worth being promoted in the production. Now it has been adopted in many tanneries.

Table 4.9 The item and the properties of the leather different technology of deliming

Item	Deliming technology with carbon dioxide	Deliming technology with ammonium sulfate
Content of calcium/(mg/L)	980.0	870.0
Content of ammonium/(mg/L)	6.0	27.0
Content of nitrogen/(mg/L)	5.0	22.0
pH	7.3	7.6
Tensile strength/(N/mm^2)	22.46	21.10
Elongation/%	62.5	50.0
Tear strength/(N/mm)	26.6	26.0
Ash content/%	4.84	4,92
Water-soluble substance/%	1.27	1.28
Content of Cr_2O_3/%	3.73	3.52
Content of grease/%	10.5	11.0

Magnesium lactate, magnesium sulfate, magnesium chloride, etc also is respectively used in the de-liming process in order to obtain the aim of reducing the content of ammonia nitrogen in the effluent. During the deliming process, magnesium salts can react with lime and form insoluble magnesium hydroxide, calcium sulfate or soluble calcium lactate, which can improve the fineness of the grain of the leather and evidently decrease the content of ammonia nitrogen of the effluent, as seen from Table 4. 10.

Table 4. 10 The content of ammonia nitrogen of the effluent of different deliming agents

Deliming agents	Content of ammonia nitrogen of the effluent/ (mg/L)	
	After bating	After tanning
Magnesium lactate	135	163
Ammonium sulphate	4871	1801

In addition, organic acid (formic acid, lactic acid, citric acid, etc) and boric acid will be used in the deliming process. The acidity of organic acid is stronger than that of ammonium salt. It may cause the pelt swelling and the quality of the leather is declining when the acid is solely used in the deliming process. Nevertheless, the acidity of boric acid is weak so that it can react with lime and form soluble calcium metaborate to achieve the aim of deliming. The technology of deliming process with boric acid is convenient and easy to be safely controlled. The properties of wet blue delimed by boric acid is equivalent to that of ammonium sulfate. But the cost of the technology of deliming process is higher than that of ammonium sulfate. In order to decrease the cost of deliming process, it can be matched with citric acid, sodium citrate.

The deliming materials without ammonium is difficult to permeate into the pelt so that the time of deliming is longer than that of ammonium salt. Therefore, the deliming technology with less ammonia is adopted by lombining the deliming materials without ammonium and ammonium salt, which can not only be easy to be operated and shorten the time of deliming, but also reduce the content of ammonia nitrogen in the effluent.

4. 3 Clean technology of chrome tanning operation

Tanning operations include pickling and tanning process. Pickling can decrease the pH of pelt, restrain the reaction of enzymes, further open up the collagen fibers and be ready for the subsequent tanning process. But sodium chloride used in the pickling process causes the pollution of chloride ion of the effluent.

4. 3. 1 Clean technology of pickling process

In recent years, many technologists studied and carried out the pickling without salt or non-pickling so that it can reduce the pollution of chloride ion and improve the absorption level of chromium.

The technology of pickling without salt is implemented by adopting no - swelling acid or its compounds to obtain the aim of pickling. During the process of pickling without salt, pelt can't swell

4-4

and the condition of the operation, cost, the quality of the finished leather is equivalent to that of traditional technology. Shan Zhihua, et al studied auxiliaries synthesized by sulfonation and condensation of aromatic hydrocarbon with one or more rings and phenol and used it in the process of pickling without salt. Results show that the technology of pickling without salt reduces more than 80% of the pollutions of chloride ion and the quality of the finished leather is equivalent to that of traditional technology. The auxiliaries used in the process of pickling without salt increase the cross-linking points for chrome tanning agent and achieve the pretanning effect so that it can improve the absorption level and rate of combination of chromium. Alois Puntener, et al studied a tanning system of Ciba. Wherein the polymer based on sulfone sulfonic acid and little formic acid are used for pickling under the condition of no salt. The technology of pickling of limed and bated bovine pelt is shown in Table 4. 11, reduces the concentration of chloride ion in the effluent from 27kg to 1kg every ton of hide and contributes to the penetration and absorption of chromium in the pelt. Compared with the shrinkage temperature of the leather used by conventional picking and chrome tanning, the in creasement of the shrinkage temperature of the leather exceeds 10℃.

Table 4. 11 The technology of pickling of limed and bated bovine pelt

Process	Chemicals	Offer/%	T/℃	t/min	pH	Remarks
Pickling	Water	50	25			
	Polymer based on sulfone sulfonic acid	2		60	4. 4	
	Polymer based on sulfone sulfonic acid	2				
	Formic acid (85%)	0. 4		60	3. 2	Check the penetration of acid

R. Palop and A. Marsal studied the effect of repression of swelling by adopting polyacid, 3, 6-naphthol disulfonic acid, p-hydroxydiphenylsulfonic acid and Retanal SCN (mixture of naphthalene sulfonic acid and naphthol sulfonic acid), respectively. Results show that the effect of Retanal SCN is the best based on the properties of the finished leather. Retanal SCN can repress the acid swelling of the pelt, and raise the shrinkage temperature of the leather when the concentration of salt solution is 2°Be. R. Palop ulteriorly studied and optimized the technology of pickling and chrome tanning with Retanal SCN and Basicromo BA (Being easy to penetrate and masking highly) when the concentration of salt solution is 0° Be. The technology of pickling of limed and bated bovine pelt is shown in Table 4. 12.

Table 4.12 The technology of pickling and tanning of limed and bated bovine pelt

Process	Chemicals	Offer/%	T/℃	t/min	pH	Remarks
Pickling	Water	60	25			
	Retanal SCN	3		30		
	Sulfuric acid(1 : 10)	0.6		30		Check the pH
Tanning	Basicromo BA	3		180		
	Chrome tanning agent (B=33%)	5		120		
	Water(50℃)	50		15		
	Plenatol HBE (Basifying agent)	0.6		480		

Wondu Legesse, Thanikaivelan, et al studied the technologies of less pickling without salt, chrome tanning on the condition of different pH and compared with conventional pickling and chrome tanning. The effect of the penetration, the absorption of chrome tanning agent, the shrinkage temperature, the physical and mechanical properties of the leather, the content of chromium (Cr_2O_3) and its distribution, etc on the condition of the system of different pickling are shown in Table 4.13, Table 4.14, and Table 4.15.

Table 4.13 The time of penetration and the absorption of chrome tanning agent on the condition of the system of different pickling

System of different pickling	pH	Time of penetration/h	Absorption level of chromium/%	T_s/℃
Sulfuric acid	8.0	5.5	95	115
	7.0	6.25	94	118
	6.0	4.5	95	>120
	5.0	4.0	94	>120
Formic acid and acetic acid	8.0	11.5	91	116
	7.0	13.0	89	117
	6.0	8.5	85	>120
	5.0	7.5	86	>120
Formic acid and oxalic acid	8.0	7.25	96	118
	7.0	8.0	95	119
	6.0	4.25	92	>120
	5.0	4.25	93	>120
Sulfuric acid	4.0	7.0	74	>120
	3.0	4.0	65	>120

Table 4.14 The physical and mechanical properties of the leather

System of different pickling	pH	Tensile strength/(N/mm)	Elongation at break/%	Tear strength/(N/mm)	Spalling strength of grain	
					Load/N	Height/mm
Sulfuric acid	8.0	18.1	59	67	196	7.86
	7.0	19.2	56	71	196	8.60
	6.0	23.1	80	97	274	10.82
	5.0	22.2	57	113	245	8.46
Formic acid and acetic acid	8.0	17.4	56	71	206	7.94
	7.0	18.3	77	61	186	10.23
	6.0	22.2	71	92	265	10.20
	5.0	23.1	72	124	382	11.44
Formic acid and oxalic acid	8.0	19.4	65	58	196	7.98
	7.0	22.0	78	72	186	9.09
	6.0	20.8	65	130	353	10.61
	5.0	22.1	66	65	255	9.34
Sulfuric acid	4.0	23.4	88	71	353	10.19
	3.0	26.3	94	93	304	9.35

Table 4.15 The content of chromium (Cr_2O_3) and its distribution in the wet blue

System of different pickling	pH	Content of chromium (Cr_2O_3)/%			Total content of chromium (Cr_2O_3)/%
		Grain layer	Middle layer	Reticular layer	
Sulfuric acid	8.0	5.59	2.31	5.30	4.51
	7.0	5.43	2.51	4.01	4.98
	6.0	5.26	4.83	5.24	5.08
	5.0	5.29	5.16	5.26	5.02
Formic acid and acetic acid	8.0	5.02	2.82	5.10	4.69
	7.0	5.10	2.78	4.60	4.56
	6.0	5.20	4.22	5.62	4.73
	5.0	4.98	4.75	4.73	4.56
Formic acid and oxalic acid	8.0	5.54	2.67	5.39	4.81
	7.0	5.04	2.85	4.35	4.57
	6.0	5.53	4.37	5.93	4.98
	5.0	5.02	4.82	4.93	4.96
Sulfuric acid	4.0	3.86	2.88	3.06	3.26
	3.0	2.60	2.51	2.75	2.75

It can be seen from Table 4. 13 that the absorption of chromium under the condition of high pH is evidently higher than that of conventional pickling and chrome tanning, the technology has clearly advantage in terms of the time of penetration and tanning, which may be the penetration and binding of chromium are simultaneously proceeding. In addition, the shrinkage temperature of the leather from the technologies of less pickling without salt and conventional pickling and chrome tanning can reach over 110℃. It can be seen from Table 4. 14 and Table 4. 15 that the technology of less pickling without salt can obtain the better effect of penetration of chromium in the pelt and the distribution of chromium is uniform at pH 5. 0 and 6. 0. But the physical and mechanical properties of the leather slightly decline than those of conventional pickling and chrome tanning. When pH is in the range of 5. 0 and 6. 0, they are similar to those of traditional methods. Therefore the technology of less pickling without salt can proceed on the condition of higher pH, simplify the operation and shorten the time of the process, improve the absorption of chromium and reduce the pollution of chloride ion and chromium. The quality of the leather is close to that of traditional methods and it is further improved by the balance of the technologies.

Luo Jianxun, et al developed a non-pickling chrome tanning technology. A macromolecular aliphatic aldehyde, PTA was synthesized and employed to pretan bated pelt before chrome tanning so that the conventional pickling process was eliminated (Figure 4. 1). It was found that PTA possesses moderate tanning property itself and favors the penetration of chrome, resulting in chrome exhaustion of 90%. Meanwhile, the performance of leather was comparable to the conventionally processed leather in terms of hydrothermal stability and property evaluations.

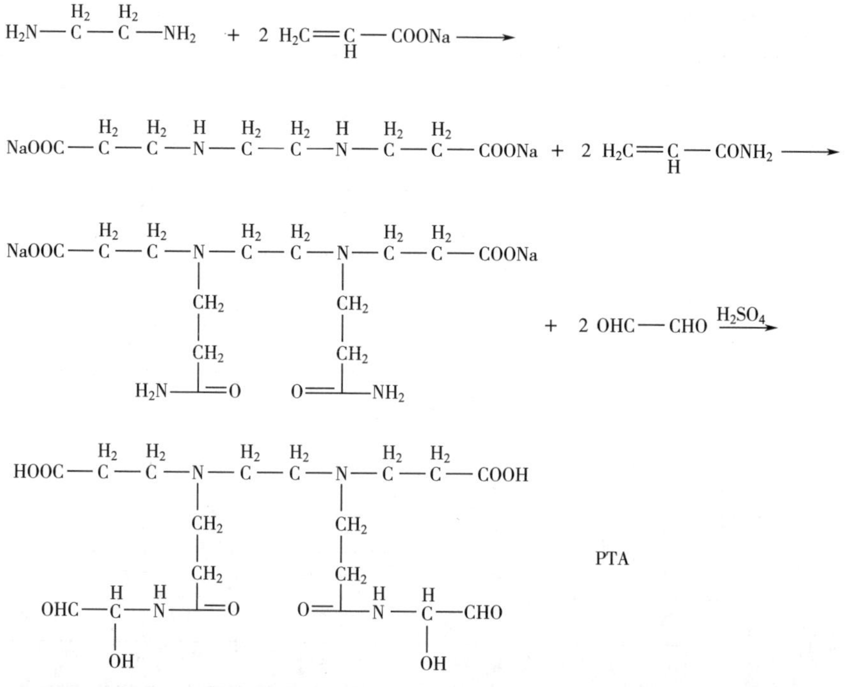

$$H_2N-CH_2-CH_2-NH_2 + 2\ H_2C=CH-COONa \longrightarrow$$

$$NaOOC-CH_2-CH_2-NH-CH_2-CH_2-NH-CH_2-CH_2-COONa + 2\ H_2C=CH-CONH_2 \longrightarrow$$

$$NaOOC-CH_2-CH_2-N(CH_2CH_2CONH_2)-CH_2-CH_2-N(CH_2CH_2CONH_2)-CH_2-CH_2-COONa + 2\ OHC-CHO \xrightarrow{H_2SO_4}$$

$$HOOC-CH_2-CH_2-N(CH_2CH_2C(=O)NH-CH(OH)-CHO)-CH_2-CH_2-N(CH_2CH_2C(=O)NH-CH(OH)-CHO)-CH_2-CH_2-COOH \quad \text{PTA}$$

Figure 4. 1 The synthesis of pretanning agent PTA

4.3.2 Clean technology of chrome tanning process

4-5

Tanning is a key process during the process of leather manufacture which changes the hide or skin into the leather. The properties of the leather directly depend on the tanning effect of tanning process. As far as tannage is concerned, the chrome tanning agents are considered the best mineral tanning agent among the tanning agents used in the leather industry because it is easy to be controlled and has good comprehensive properties such as high hydrothermal stability, softness, etc. Therefore, most of the tanneries in the world have adopted chrome tannage. The modern system of leather manufacture based on chrome tannage has been formed after more than 100 years of development. However, the absorption level of chromium is only 65%–75% during traditional chrome tanning so that the tannery effluent contains high salinity, chromium and other wastes containing chromium are produced which have a serious environmental impact. Therefore clean tanning is one of the most important contents in the cleaner production of leather manufacture. How to reduce or abolish the disadvantages of the chromium tanning agent is always an important topic in the leather industry. The key problem is to improve the absorption level in order to overcome the absorption and reaction between collagen and chrome tanning agents. The traditional chrome tannage that involved the use of sodium chloride, acid and chromium is one of the main origins of salt and chromium pollution.

Many technologists have obtained results in the field of chrome tannage without pickle, high ethau-stion chrome tannage, chrome-reduced tannage and chrome-free tannage. Based on the needs and trends of leather products and markets, chrome-tanned leather and non-chrome tanned leather will coexist for a long time. Therefore, in order to meet the demands of markets and environmental protection, the one of best ways is to reduce the dosage of acid, chrome tanning agents and improve the absorption level of chromium in the tanning process, so chrome tannage without pickle, chrome tannage with high absorption and chrome-reduced tannage are the cleaner chrome tanning technologies.

Researches on the mechanism of chrome tannage show that the main reaction is coordination reaction between the carboxyl group of collagen and chromium complexes. Therefore the rule of chrome tanning is firstly the penetration and then the reaction. One of the aims of pickling process is to restrain dissociation of the carboxyl group of collagen and inhibit the tendency of the hydrolysis of chromium salts, association reaction and molecular enlargement. But the pickling process causes the pollution of chloride ions and acid as previously mentioned. In order to overcome the shortcomings of traditional pickling and chrome tanning, the technology of chrome tanning without pickle is proposed by some chemists in the field of leather manufacture, which is to make a new chrome tanning agent that can successfully penetrate into the inner layer of the leather and achieve the ideal pH ranges by itself. Chen Wuyong, et al studied and produced a chrome tanning agent without pickle (C-2000, Cr_2O_3 21%±1%). Results of applications of C-2000 show that C-2000 can successfully penetrate into the inner layer of the leather on the condition of high pH. The leather tanned by C-2000 has good properties such as fullness, high shrinkage temperature, high chromium content, evenly distribution, etc. P. Thanikaivelan, et al invented three chrome tanning agents without pickle which is prepared by the reaction between the sulphonate of aromatic compounds AHC, carboxylic acid compounds PCA and organic acid respectively and a chrome tanning agent BCS. Results of controlled experiments show that the high shrinkage temperature of the leathers tanned by three chrome tanning agents achieves more than 110℃, the chromium content in the leather is higher than that of conventional chrome tanning and the absorption of chromium obtains over 90%, etc. Therefore the technology of chrome tanning without pickling and corresponding chrome tanning agent without pickle are feasible by comprehensive considering the cost, the properties of the leather, the absorption of chromium, etc.

4-6

4.3.2.1 Optimizing parameters during the chrome tanning

In order to improve the absorption of chromium during the chrome tanning process, many chrome tannages with high absorption are studied and applied to the tanning process. During the tanning process, the mechanical action, the liquid ratio, the amount of chrome tanning agents, the sorts and the amount of masking agents, pH, temperature and time of tanning, etc are important parameters. Therefore the absorption of chromium can be improved and the content of Cr_2O_3 in the effluent is reduced by optimizing these parameters.

4.3.2.2 High exhaustion of chromium

In addition, many chemists in the field of leather manufacture developed novel auxiliaries and improved the technology of tanning in order to achieve the high exhaustion of chromium in the production of leather. These novel auxiliaries include aliphatic dicarboxylate, glyoxalic acid, oxazolidine acid, polymers with more carboxyl groups, amino groups, hydroxyl groups, etc. The first three auxiliaries are classified as materials with small molecular weight.

J. Gregori, et al propose that aliphatic dicarboxylate with 4-6 or 8-13 carbon atoms can play the role of cross-linking agent when it is applied in the process of chrome tanning. The carboxyl groups of aliphatic dicarboxylate can link the chromium complexes combined with hide fibers by a single point and form cross-linked bonds of multipoint combination so as to improve the shrinkage temperature of the leather, the absorption and fixation of chromium. Thus the absorption of chromium can achieve 85% and the content of Cr_2O_3 in the effluent is 1g/L or so. But the distribution of Cr_2O_3 in the leather is poorer than that of the traditional chrome tanning process and the grain, color of the leather is coarse, green. Li Guiju, et al prove the results by tanning experiments adopting phthalic anhydride as an auxiliary of chrome tanning. The formulation of the technology is shown in Table 4.16.

Table 4.16 The formulation of the technology of tanning

Process	Chemicals	Offer/%	T/℃	t/min	pH	Remarks
Pickling	Pickling liquid	100	18-22		2.8	
Tanning	Chrome tanning agent	2		90		
	Phthalic anhydride	1.5		60		
	water	100	40	120		
Basification	Sodium bicarbonate	1.5		120	~5.0	
Stopping drum overnight, running 30min next day and horse up						

Because glyoxalic acid contains the aldehyde group and carboxyl group, it not only has tanning properties, but also can increase the cross-linking points of chromium (Ⅲ). Glyoxalic acid can irreversibly react with amino groups of the collagen so as to increase the carboxyl groups in the collagen fibers. The reactions among glyoxalic acid, collagen and chrome tanning agents are shown in Figure 4.2. Thus it contributes to increasing the fixation of chromium, improving the shrinkage temperature and reducing the content of Cr_2O_3 in the effluent. During the pickling process, glyoxalic acid can replace part of sulfuric acid and formic acid so as to reduce the amount of acid and sodium chloride. In addition, it is woted that glyoxalic acid has masking properties which is adverse to the hydrolysis, olation and cross-linking of chromium complex. But the cost of glyoxalic acid is high so that it limits the application in the production of leather manufacture. Therefore, in order to decrease the cost of production and achieve the high absorption of chromium, Li Guoying, et al, Fan Haojun, et al and Bai Yunxiang, et al developed aldonic acid tanning aid LL-Ⅰ, aldonic acid tanning agent AA and SYY respectively. Aldonic acid tanning aid LL-Ⅰ is prepared by the Michael reaction of aldehyde, ester under the condition of alkali acting as a catalyzer. Aldonic acid tanning agent AA is prepared by condensation reaction between organic acid with active hydrogen and reaction products of glutaraldehyde, and formaldehyde, which can promote the absorption and cross-linking of chromium, the finished leather is full and has good mechanical properties. Aldonic acid tanning aid SYY has many aldehyde groups, carboxyl groups and hydroxyl groups which can react with the collagen of pelt.

Oxazolidine acid is also an auxiliary with small molecular weight which has MOCA with a single ring and OXD-Ⅰ with two rings and can improve the absorption of chromium. Their structures are follow as Figure 4.2. Wang Hongru, et al develop an oxazolidine acid MOCA with the structure of the ring of oxazolidine and carboxyl groups synthesized by threonine and formaldehyde. The applied formulation and results of MOCA in the chrome tanning process are shown in Table 4.17, Table 4.18, respectively. It can be seen from Table 4.17 that MOCA can improve the the absorption of chromium and reduce the content of Cr_2O_3 in the effluent during the conventional chrome tanning process. Fan Haojun, et al develop and produce an oxazolidine acid OXD-Ⅰ whose applied formulation and results are shown in Table 4.19, Table 4.20. OXD-Ⅰ can produce highly active azoxymethyl groups, hydroxyl groups by hydrolysis and opening loop. Then the azoxymethyl groups react with the amino groups of the side chain of collagen and form stable covalent bonds. The hydroxyl groups also can form multi-point hydrogen bonds with peptide bonds, carboxyl groups, amino groups, hydrogen groups, etc. These actions form synergism so that OXD-Ⅰ can be fixed on the collagen fibers. The carboxyl groups on the OXD-Ⅰ and side chain of collagen can coordinate with chrome complex and form irreversible binding by coordinating cross-linking, single point, cyclic chelation, etc when chrome tanning agent is added into the system, which is shown in Figure 4.2. Thus the irreversible coordinating cross-linking improves the fixation of chromium and reduces the content of Cr_2O_3 in the effluent.

Table 4.17 The applied formulation of MOCA in the chrome tanning process

Process	Chemicals	Offer/%	T/℃	t/min	pH	Remarks
Pretanning and pickling	Water	30	25-30			
	MOCA	1.0-5.0		480		Diluted with water
	Formic acid	0.5-1.0		40-60	3.5-3.7	
Tanning	Chrome tanning agent	6		90		
Basification	Sodium bicarbonate	1.6		120±30	4.0-4.2	
	Water	120	40	360		
Stopping drum overnight, running 30min next day, horse up and testing T_s, pH and the content of Cr_2O_3 in the effluent.						

Table 4.18 Applied results of MOCA in the chrome tanning process

Projects	Numerical value				
Amount of MOCA/%	0	1.0	2.0	3.0	4.0
T_s of Leather pretanned/℃	52	61	63	67	69
T_s of Leather tanned by Chromium (Ⅲ)/℃	97	100	104	107	109
Content of Cr_2O_3 in the effluent/(g/L)	2.151	2.015	1.364	0.628	0.516
Absorption of chromium/%	79.32	80.63	86.88	93.96	95.04

Table 4.19 The applied formulation of OXD-I in the chrome tanning process

Process	Chemicals	Offer/%	T/℃	t/min	pH	Remarks
Depickling	Water	30	18-22			
	Sodium chloride	4		30		
	Sodium bicarbonate	0.5		60	3.0-3.5	
Tanning	OXD-I	2		90		Diluted with water
	Chrome tanning agent	5		180		
Basification	Sodium bicarbonate	1.2-1.5		120+30	4.0	
	Water	150		120		
Stopping drum overnight, running 30min next day, horse up and testing T_s, pH and the content of Cr_2O_3 in the effluent.						

Table 4.20 Applied results of OXD-I in the chrome tanning process

Projects	Numerical value				
Amount of OXD-I/%	0	1.0	2.0	3.0	5.0
Content of Cr_2O_3 in the effluent/(g/L)	1.71	0.68	0.18	0.17	0.12
Absorption of chromium/%	71.5	85.8	96.0	96.5	97.5
T_s of Leather tanned by Chromium (Ⅲ)/℃	106	110	112	113	114

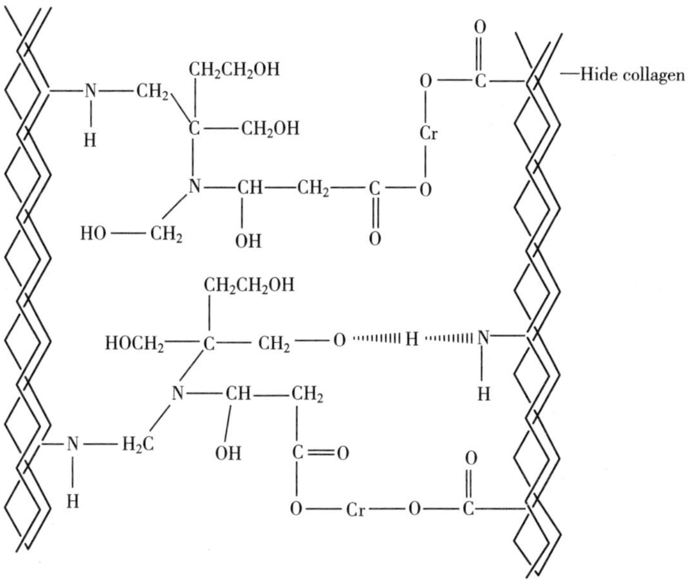

Figure 4.2 Diagram of reaction among chromium (Ⅲ), OXD-I and hide collage

It can be seen from Table 4.20 that the absorption of chromium is more than 96%, the content of Cr_2O_3 in the effluent is less than 200mg/L and the shrinkage temperature of the leather exceeds 100℃ when the amount of OXD-Ⅰ is in the range of 2% and 3%. The leather is full and has excellent dyeing properties and physical-mechanical properties.

In order to further improve the effect of chrome tannages with high absorption, Duan Zhenji prepares a polymer with more carboxyl groups, amino groups, hydroxyl groups (Being named as PCPA). Zeng Weiyong, et al develop PCPA-Ⅱ on the basis of PCPA. Fan Haojun, et al develop ECPA with carboxyl groups, amino groups, aldehyde groups, hydroxyl groups, benzene rings, etc which are synthesized by free radical co-polymerization among 2-aldehydeethyl acrylate, acrylic acid and other vinyl monomers. Results of applied experiments of PCPA, PCPA-Ⅱ, and ECPA show that the content of Cr_2O_3 in the effluent less than 200mg/L and the shrinkage temperature of the leather exceeds 110℃. In addition, Wang Xuechuan, et al synthesize a hyperbranched polymer acting as an auxiliary of chrome tanning, which provides a wide development direction for developing the polymer auxiliary of chrome tanning.

4.3.2.3 Chrome-reduced tannage

Chrome-reduced tannage is also one of the clean production of leather manufacture. Concerning existing chrome-reduced tannage, one way is to adopt some organic tanning agent or other mineral tanning agent to achieve combination tanning with a chrome tanning agent. These organic tanning agents include vegetable tannin, modified glutaraldehyde, amino resin, etc. In the combination tannage of vegetable tannin and chromium, vegetable tannin can evenly distributed in the pelt when its amount exceeds 15%. However, the leather has a strong vegetable tanning character. Therefore in order to eliminate the vegetable tanning character of the finished leather and improve the shrinkage temperature of the leather, Shi Bi, et al study the modification of vegetable tannin so as to reduce the vegetable tannin and ensure the fast, even penetration of vegetable tannin in the pelt. The oxidative, degradation and modification of valonia extract is used in the combination tannage of vegetable tannin and chromium. The applied formulation, and results of the combination tannage of vegetable tannin and chromium are shown in Table 4. 21, Table 4. 22. It can be seen from Table 4. 21 that the shrinkage temperature of the leather exceeds 100℃ when the usage of Cr_2O_3 is 1%. Compared with the shrinkage temperature, grain and state and rate of thickening of the leather on the condition of different amounts of modified valonia extract and chrome tanning agent, the optimum technology is the combination tannage of 5%-7% vegetable tannin and 1.0%-1.5% chromium (Cr_2O_3). Thus the amounts of chrome tanning agent decrease 25%-50% than that of conventional chrome tanning.

Table 4. 21 Applied formulation of the combination tannage of vegetable tannin and chromium

Process	Chemicals	Offer/%	T/℃	t/min	pH	Remarks
Pretanning	Water	50	18-22			
	Sodium chloride	5		30		
	Fat-liquoring agent/ Assisted syntan	2		60		
	Modified extract	5/7/10		90	5.0	Check the penetration
	Water (40℃)	50		120		
	Formic acid	0.3-0.5			3.1-3.2	
Tanning	Chrome tanning agent (B=35%, Cr_2O_3 21%)	0.5/1.0/ 1.5/2.0		120		Amount of chrome tanning agent depends on the content of Cr_2O_3
Basification	Sodium bicarbonate	1.2-1.5		120+30	3.8-4.0	
	Water	100	40	120		

Stopping drum overnight, running 30min next day, horse up and testing T_s, pH and the content of Cr_2O_3 in the effluent.

Table 4. 22 Applied results of the combination tannage of vegetable tannin and chromium

Amount of Chromium (Based on Cr_2O_3)		0.5%	1.0%	1.5%	2.0%
5% Amount of modified valonia extract	T_s of the leather/℃	95	110	116	121
	Grain and state of the leather	Fine and soft	Fine and soft	Fine and soft	Fine and soft
	Rate of thicken /%	40.3	48.2	57.6	62.9
7% Amount of modified valonia extract	T_s of the leather/℃	94	106	120	123
	Grain and state of the leather	Fine and soft	Fine and soft	Fine and soft	Fine and soft
	Rate of thickening/%	50.4	54.6	64.6	67.8
10% Amount of modified valonia extract	T_s of the leather/℃	92	103	117	129
	Grain and state of the leather	Fine and soft	Fine and soft	Coarse and soft	Coarse and hard
	Rate of thicken /%	56	56.8	73.6	70.8

Shi Bi, et al study the deep sulfitation reaction of larch tannin extract in order to reduce the molecular weight of larch tannin extract or open the arsenic ring of tannin and improve the solubility of the extract, decrease the astringency of the extract by sulfonic acid group. The applied formulation and results of the combination tannage of modified larch tannin extract and chromium is similar to that of the combination tannage of modified valonia extract and chromium. The optimum technology is the combination tannage of 5%–10% vegetable tannin and 1.0% chromium (Cr_2O_3). Thus the amount of chrome tanning agent decrease 50% than that of conventional chrome tanning.

Some researchers also study the combination tannage of other organic tanning agents and chromium. Shi Bi, et al studied the chrome-reduced tannage of goatskin garment leather by adopting modified glutaraldehyde with chromium. The applied formulation, results of the combination tannage of modified glutaraldehyde and chromium is shown in Table 4.23, Table 4.24. It can be seen from Table 4.24 that the optimal consumption of modified glutaraldehyde is at the range of 6% and 8% when the goatskin is tanned by the combination tannage between modified glutaraldehyde and 0.5% Cr_2O_3. The shrinkage temperature of the leather reaches 95℃ and the tannage is good for subsequent retanning, dyeing and fat-liquoring. In addition, Qiang Xihuai, et al studied the amino resin-aldehyde-chromium tannage. Results show that the tannage is better than the conventional chrome tannage when the consumption of amino resin, aldehyde, chromium is 2%, 1% and 4% respectively. The leather is fine, full, soft, light color and reduces 50% the amount of chromium. The content of Cr_2O_3 in the effluent is 390mg/L. Modified starch is also used to pretan the pelt in order to improve the shrinkage temperature of the leather and reduce the content of Cr_2O_3 in the effluent.

Table 4.23 Applied formulation of the combination tannage of modified glutaraldehyde and chromium

Process	Chemicals	Offer/%	T/℃	t/min	pH	Remarks
Pretreatment	Pickling liquid	100	18–22			
	Sulfited fish oil	2		45		
Tanning	Pickling liquid	50				
	Modified glutaraldehyde	6		90		
	Sodium acetate	3–4		3×20+30	4.5	
Stopping drum overnight, running 30min next day						
Retanning	Liquid ofaldehyde-tanning	100				
	Chrome tanning liquid (B=38%)	0.5		240	5.0	Amount of chrome tanning liquid is based on the content of Cr_2O_3
Basification	Sodium bicarbonate	1.0–2.0		4×15+30	3.8–4.0	
Stopping drum for over night, horse up						

Table 4. 24 Applied results of the combination tannage of modified glutaraldehyde and chromium

Projects	Numercial value					
Amount of modified glutaraldehyde/%	2	4	6	8	10	12
T_s of leather tanned by modified glutaraldehyde/℃	69. 0	70. 0	73. 0	79. 0	80. 0	83. 0
T_s of leather retanned by 0. 5% Cr_2O_3/℃	89. 0	92. 0	94	95	95	96

The other way is to tan the pelt with a new chrome tanning agent modified by other mineral tanning agents. In recent years, in order to improve the effect of chrome-reduced tanning, the chrome-reduced tannage based on wet-white has been studied, that is, the pelt is pretanned with an aluminium tanning agent, or aldehyde tanning agent to obtain the wet-white, then the shaved wet-white is tanned with a chrome tanning agent. However, the appearance, softness and other properties of the chrome-reduced tanned leathers are lower than those of a traditional chrome tannage, thus, chrome-reduced tannage needs further study.

Luo Jianxun, et al developed a non-pickle, chrome-reduced tanning technology. The novel chrome-free agent SL can be directly employed to tan bated bovine hide producing wet-white (The synthesis of the novel chrome-free agent SL is shown in Figure 4. 3).

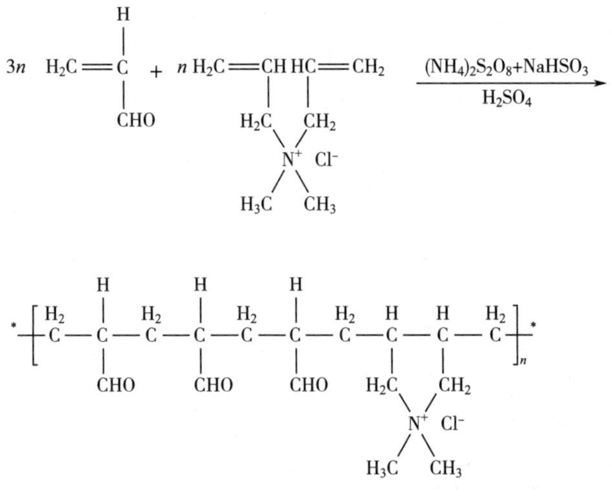

$$3n\ H_2C{=}C(H)(CHO) + n\ H_2C{=}CH\,HC{=}CH_2\ [H_2C,\ CH_2 \rightarrow N^+(H_3C)(CH_3)\ Cl^-] \xrightarrow[H_2SO_4]{(NH_4)_2S_2O_8+NaHSO_3}$$

$$\text{*}\!-\!\left[\overset{H_2}{C}-\underset{CHO}{\overset{H}{C}}-\overset{H_2}{C}-\underset{CHO}{\overset{H}{C}}-\overset{H_2}{C}-\underset{CHO}{\overset{H}{C}}-\overset{H_2}{C}-\underset{H_2C}{\overset{H}{C}}-\underset{CH_2}{\overset{H}{C}}-\overset{H_2}{C}\right]_n\!-\!\text{*}\quad (H_2C,\ CH_2 \rightarrow N^+(H_3C)(CH_3)\ Cl^-)$$

Figure 4. 3 The synthesis of the novel chrome-free agent SL

The shaved wet-white was pretreated by a poly-carboxylate auxiliary agent PAA and tanned by chrome powder (Reaction scheme PAA is shown in Figure 4. 4).

It was found that the shrinkage temperature of the wet-white tanned by SL reached over 80℃, the optimal consumption of poly-carboxylate auxiliary agent was 1. 5%-2% (wt) based on the weight of the shaved wet-white, the better chrome-reduced tanning conditions were that the wet-white was tanned with 3-4wt% chromium powder for 150-180 minutes at room temperature when the initial pH value was 3. 0-3. 5. The next processes were the same as those of a traditional chrome tannage. Meanwhile,

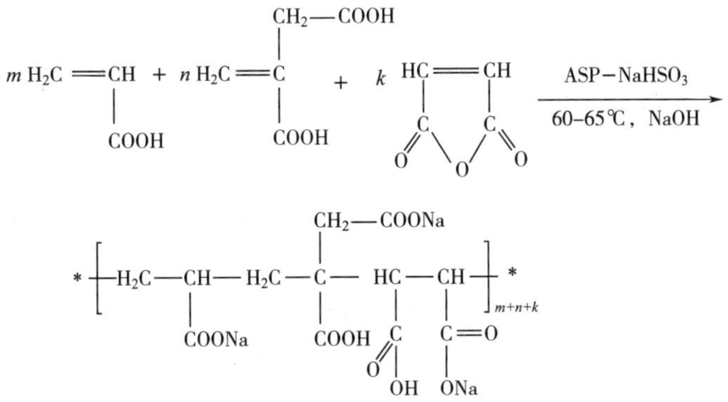

$$m\,H_2C{=}CH(COOH) + n\,H_2C{=}C(CH_2{-}COOH)(COOH) + k\,HC{=}CH\ (\text{maleic anhydride}) \xrightarrow[60\text{-}65^{\circ}C,\ NaOH]{ASP{-}NaHSO_3} *{-}[H_2C{-}CH(COONa){-}H_2C{-}C(CH_2{-}COONa)(COOH){-}HC(COOH){-}CH(COONa)]_{m+n+k}{-}*$$

Figure 4. 4 Schematic diagram of the synthetic reaction

the shrinkage temperature of the leather tanned by the chrome-reduced tannage reached more than 95℃, the absorption of chromium was 96%, and the content of Cr_2O_3 in the effluent was under 200 mg/L. For the chrome-reduced tanned leather, the absorption of dyestuff and fat-liquor reached 99.5% and 82.5%, respectively. Compared with the traditional chrome tanning process, not only was the conventional pickling process eliminated, but the process was shortened and reduced the pollution of sodium chloride. The process can reduce by 50% the consumption of chrome powder, improve the absorption of chromium and reduce the content of Cr_2O_3 in effluent.

Other mineral tanning agents include aluminium tanning agent, rare earth, iron tanning agent, zinc tanning agent, etc. Although the tanning properties of these tanning agents are weaker than that of chrome tanning agent, they can be combined with chromium in order to reduce the amount of chromium.

For the combination tannage of aluminium tanning agent and chromium, A. D. Covington did a lot of research and put forward that little aluminium tanning agent can promote the absorption of chromium, reduce the amount of chromium, cut the cost of chrome tanning, etc. The color of the leather is lighter than that of conventional chrome tanning. But the effects arenot significant in terms of decreasing the consumption of chromium tanning agent and reducing the pollution of chromium. Sreeram, et al studied an applied technology of a syntan containing aluminium Alutan in chrome tanning. Results show that the absorption of chromium can obtain optimal effects when Alutan is simultaneously added into the drum with chromium salt. The absorption of chromium is drastically improved when the pelt is tanned by 1.5% Alutan and 5% chromium tanning agent. The shrinkage temperature, physical-mechanical properties and state of the leather are equal to or superior to that of conventional chrome tanning so as to achieve the aim of reducing the amount of chromium tanning agent. In addition, chromium salt and aluminic acid can form chromium-aluminium multi-nuclear complex so as to obtain the aims of decreasing the consumption of chromium tanning agent, increasing the absorption of chromium, reducing the pollution of chromium and the cost of production.

On the whole, the chromium-aluminium tannage can decrease the consumption of chromium tanning agents, increase the absorption of chromium, and reduce the pollution of chromium and the cost of production. The physical-mechanical properties and state of the leather are similar to that of conventional chrome tanning.

Rare earth has certain tanning properties which is weaker than that of chromium (Ⅲ) , zirconium (Ⅳ) , titanium (Ⅳ) , etc. The shrinkage temperature of the leather tanned by rare earth is 63℃ or so and the leather is not rinsed. Zhang Mingrang, et al found that rare earth can be matched with chromium so as to show the advantage of combination tanning. When rare earth acts as an auxiliary in chrome tanning, it can markedly increase the tanning effect including improving the quality, grade and gain of the finished leather, increasing the absorption of chromium, reducing the consumption of chromium salt and decreasing the content of Cr_2O_3 in the effluent, etc. In the initial stage of chrome tanning, rare earth can promote the increase of anionic, neutral chromium complex which benefits the penetration of chromium in the pelt. However, in the later stage of chrome tanning, when the pH of the float rises by basification, rare earth can also promote to turn anionic, neutral, low positive chromium complex into high positive chromium complex which is in favor of cross-linking between chromium (Ⅲ) and collagen. The technology of chrome tanning assisted by rare earth can save 40% - 50% chromium salt.

Many researchers also studied the combination tannage of chromium with iron tanning agent, zinc tanning agent, etc. P. Thanikaivelan, et al develop a novel chromium-iron tanning agent whose content of Cr_2O_3 and Fe_2O_3 is 14. 2% and 12. 8%, respectively. When the amount of the novel chromium-iron tanning agent is 1. 5% (Calculated by metallic oxide) , the shrinkage temperature of the leather, the absorption of chromium salt and iron salt can achieve 117℃, above 90% respectively. The content of Cr_2O_3 in the effluent is 300mg/L. The properties of the leather compare to that of the leather tanned by chromium (Ⅲ). But the mechanical properties of the leather slightly decline after the leather is placed for one year. R Karthikeyan bulid a set of standard technology of combination tanning between chromium and iron tanning agent. The quality of the leather tanned by the standard technology is excellent and the absorption of chromium salt and iron salt exceeds 95%. Cheng Fengxia, et al prepare a chromium - iron tanning agent synthesized by chrome tanned leather shavings, ferrous sulfate, sodium chromate, sodium citrate, sodium tartrate and study the integrated feasibility of combination tanning of a chrome tanning agent, iron tanning agent, vegetable tannin and dyeing, that is the pigskin pickle is tanned 1. 5% the chromium-iron tanning agent. Then it is re-tanned with 5%-8% vegetable tannin. The shrinkage temperature, physical-mechanical and dry/wet rub properties of the leather can meet the needs of the market. The combination tanning of chrome tanning agent, iron tanning agent, vegetable tannin is a technology of clean chrome tanning which can not only achieve the chrome-reduce tanning, but also save the dyestuff or dyeing process. In addition, B. Madhan, et al prepare a chromium-zinc complex tanning agent which can improve the absorption of chromium and reduce the content of Cr_2O_3 in the effluent.

4.4 Clean technology of chrome-free tanning process

As we know, leather-making is one of the most important approaches to utilizing animal hide or skin massively and effectively. Because of factors such as the pelt's low exhaustion extent of Cr (Ⅲ) in the traditional leather-making technology and dispose difficulty of waste chrome-crust, etc, which will lead to waste the natural resources and pollute the environment, in - corresponding with the new development philosophy concept of innovative, coordinated, green, open and shared development. Therefore, chrome tanning system is facing the challenge of environmental protection.

In order to get rid of the pollution of Cr (Ⅲ) and Cr (Ⅵ) from the process of chrome tanning and take the place of chrome tanning agent, chrome-free tanning technology has been developed as an effective way to reduce the pollution from chrome tanning. The tanning process does not produce any chrome-containing wastes during shaving so that it can be used for the production of chrome-free leather or chrome tanning leather. The wet white is required to have certain hydrothermal stability, be stored for a long time under the condition of wet or dry state and to be easy to be processed by mechanical actions such as splitting, shaving, buffing, etc. Therefore many researches studied non-chrome tanning agents and their tannages in recent years.

4.4.1 Tanning effect of single chrome-free tanning agent

Wet white is made by chrome-free tanning agent. These chrome-free tanning agents are classified as chrome-free inorganic tanning agents and organic tanning agents. They have different tanning properties.

4.4.1.1 Tanning with single inorganic tanning agents

The chrome-free inorganic tanning agents mainly include non-chrome metal tanning agents, silicon compounds, etc. Aluminium tanning agent, zirconium tanning agent, titanium tanning agent, etc can be used to tan the pelt and obtain the wet white.

4-7

As previously mentioned, the shrinkage temperature of the leather tanned by aluminium tanning agent, zirconium tanning agent, titanium tanning agent is 75℃, 80℃, 80℃ or so respectively. Simultaneously the leather tanned by aluminium tanning agent is fine, white and has good properties of wringing, splitting and shaving. But the leather is thin and hard. Zhang Tingyou, et al prepared a special aluminium tanning agent which has good tanning effect. The characteristic features of aluminium salts are available in plenty, cheaper, and have less impact on the environment. So aluminium salts are known to be a potential tanning agent for producing leather.

Zirconium salt may cause precipitation when the pH of the solution is in the range of 2 and 3. A. Sundarrajan, et al raise the pH of precipitation of the zirconium tanning agent by adopting zirconium oxychloride and citric acid. The shrinkage temperature of the leather can obtain 95℃. In addition, K. J. Sreeram, et al develop a zirconium tanning agent Organozir whose pH of precipitation is 5.0 ± 0.5 by coordinating between a polymer compound and zirconium salt. The technology of Organozir is similar to chrome tannage. The shrinkage temperature of the leather tanned by Organozir can obtain 84℃ or so. The properties of the leather are similar to those of the chrome tanning leather.

As far as a titanium tanning agent is concerned, Wu Xingchi studied the characteristics of masking of the titanium tanning liquid and its technology. The pH of the titanium tanning agent solution is about 0. 5 – 1. 0, because it is easy to be hydrolyzed. The titanium tanning agent solution turns into a turbid state when the pH of the solution is 1. 7. Therefore in order to control the hydrolysis of Titanium and improve its capacity of alkali resistance, oxalic acid, sodium tartrate, sodium citrate, etc act as a masking agent to be used in the titanium tanning agent and form a stable titanium complex. When the optimized mole ratio of titanium and sodium citrate is 1 : 0. 2 and the pH of the solution is adjusted to pH 6. 0 or so, the solution is still clear. Peng Biyu, et al expound on the cause of different tanning properties of several metal salt from the perspective of coordination chemistry and tanning chemistry, put forward that the tanning properties of titanium should be higher than that of zirconium(Ⅳ), aluminium(Ⅲ), iron(Ⅲ), etc and is second only to that of chromium(Ⅲ). Titaniumin salt is considered as an ideal substitute for chromium salt in terms of the resource, comprehensive properties, toxicity, etc of titanium salt. The leather tanned by titaniumin salt is white and has high hydrothermal stability and good mechanical properties.

Smit & zoon Co. Ltd developed Zeology which is a novel chrome-free tanning agent with high tanning and environment – friendly performance. The active ingredient of Zeology is similar to the earth and is prone to biological degradation. The tannage of Zeology is almost the same as chrome tannage so as to accord with existing operational habits. The tanning formulation of Zeology is shown in Table 4. 25. The shrinkage temperature of the wet white tanned by Zeology is in the range of between 70℃ and 80℃. The absorption of Zeology can obtain more than 92%. The wet white is easy to be shaved, buffered and is suitable for the production of shoe upper leather, garment leather, furniture leather, etc. In addition, the wet white has excellent flame resistance, good storage resistance, and gain of leather.

4–8

Video of Zeology

Table 4. 25 The tanning formulation of zeology based on the bated pelt

Process	Chemicals	Offer/%	T/℃	t/min	pH	Remarks
Washing	Water	100	18–22	10		Drain
Pickle	Water	50	18–22		7. 0	
	SYNTHOL EM336	0. 5				
	Sodium chloride	8		10		
	Formic acid	2		4×10		Diluted with water by 1 : 10
	Sulfuric acid	0. 6		3×10+150	2. 8	Diluted with water by 1 : 10
	Citric acid	0. 3		30		
Tanning	Zeology	3		30		
	SYNTHOL EM336			300		

Continued Table

Process	Chemicals	Offer/%	T/℃	t/min	pH	Remarks
Basification	Sodium bicarbonate	1-1.4		6×10+90	4.75	
	Water (50℃)	100	35	300		
Stopping drum overnight, horse up.						

Cromogenia-Units, Barcelona, Spain, also provides a zeolite-based tanning more sustainable, aldehyde-free, heavy metal-free processes. Zeolite-based tanning allows obtaining processes with less pollutant load (chrome (Ⅲ) and glutaraldehyde) as well as very full, very white and very highly biodegradable leathers. Zeolites are minerals, specifically porous aluminosilicates, used as absorbents, catalyzers or in water treatment, among others. These minerals are formed by SiO_4 and AlO_4 tetrahedra joined by shared oxygen atoms. There are over 200 types of zeolite, depending on tetrahedron joining:

Zeolites can be of natural or synthetic origin. Natural zeolites are obtained from sedimentary, igneous, and metamorphic rocks. Synthetic zeolites are obtained from silica-alumina gel crystallization.

Zeolite interacts with collagen in a similar way to chrome -both of them are metals-but the fact that, in general, aluminum-based products have a weaker interaction with collagen than chrome compounds, should be taken into account. Hence, less stable leathers are obtained with aluminum tanning, and a $T_s \approx 75$℃ can be reached. Because zeolite has no basicity, interaction with collagen chains occurs basically through electrostatic interactions of the carboxylate chains of glutamic and aspartic acid, and also through hydrogen bonds of the hydroxyl groups of zeolite. Zeolite structure is shown in Figure 4.5. Tanning with zeolites offers several advantages over tanning with chrome:

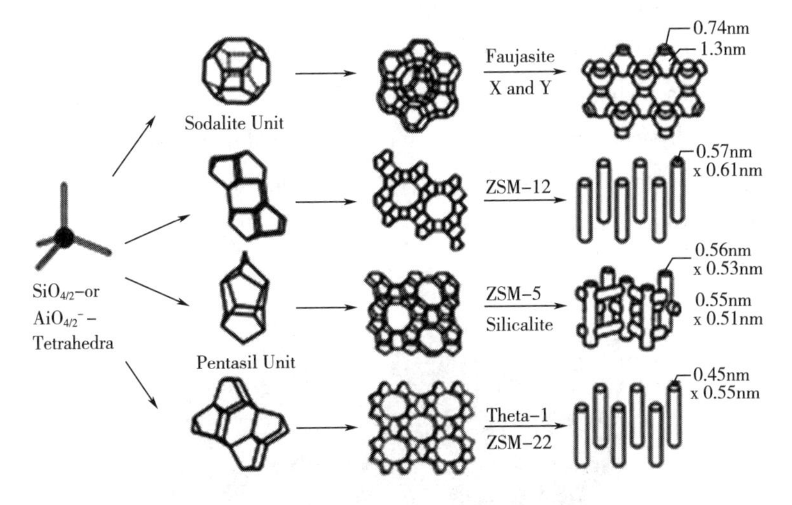

Figure 4.5 Examples of zeolite structures

- Chrome-free pelt residues
- Chrome-free wastewater and water treatment sludge
- 100% biodegradable tanned leathers

For tanning, Zeolites are dispersed at acidic pHs and therefore the penetration pH must be≤3. 5. The recommended percentage of product is 6% and 8% for cattle and lamb, respectively. It is recommended that the product be added in two takes to ensure complete exhaustion. Basification with magnesium oxide together with sodium bicarbonate is recommended to ensure complete exhaustion. Tanning with zeolites provides very white, very full, highly biodegradable leathers with $T_s \approx 75℃$.

The leathers tanned with Zeolites are cationic. Therefore, good neutralization/ anionization, preferably with a naphthalene derivative is required. Sulfone-and phenol-based synthetic retanning agents, as well as vegetable extracts, provide high fullness. Conversely, acrylic resins sometimes harden the leather and may shrink it. Accordingly, these should be added with caution. Lecithin-and fish-based natural fats provide a high degree of softness. Their addition together with a synthetic fatliquoring agent is recommended to improve penetration. Kemira Co. Ltd developed a versatile system of tanning pelt without chromium (Ⅲ), acet aldehyde and aldehyde precursor, and produced TANFOR™ T series tanning agents and its tanning system. The main active component of TANFOR™ T is a condensation of aluminum, silicon and special organic acids and is absolutely safe for human beings and the environment. Qiang Xihuai, et al, and Li Qiao ping, et al separately investigated the chemical properties and the corresponding tannage of TANFOR™ T by tanning the pickled bovine hide or goatskin (Figure 4. 6, Figure 4. 7). Results show that TANFOR™ T exhibited good tanning performances, the tanning mechanism and the operation mode of TANFOR™ T are similar to chrome tannage. The optimal tanning techniques are as follows: the pH value of pickled hide is 2. 5-3. 5, the consumption of this tanning agent is about 10% (based on the weight of the pickled goatskin), and tanning for 3h at room temperature, the penetration of TANFOR™ T in the pelt can be tested by Catechol violet reagent, then the pH value of the float is adjusted to about 5. 0, 100% hot water is added to adjust the temperature to 35-40℃, and finally tanning for 5h. The resultant wet-white goatskin leather possesses high shrinkage temperature (70-75℃), light color, good handle, excellent light fastness and good physical and mechanical properties. In addition, the leather shows an outstanding affinity for conventional re-tanning agent, dyestuff, fat-liquor and other anionic materials. After trial and wide application, the tanning agent is a novel chrome-free and ecologically friendly product and has broad application prospects.

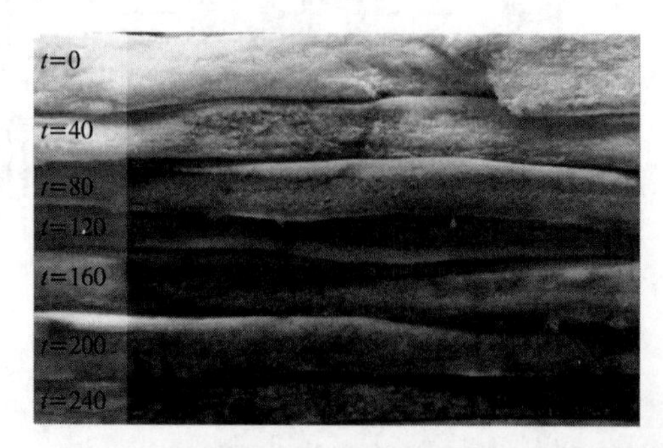

Figure 4. 6 The penetration of TANFOR™ T in the pelt when the pH of the float is 3. 0 as the time rises

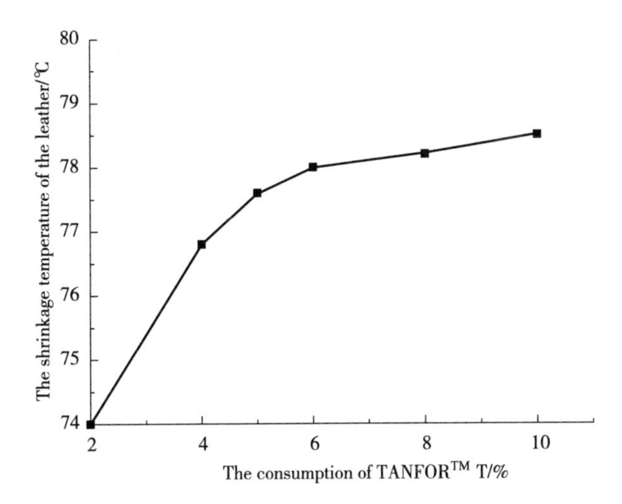

Figure 4.7 Relations between the consumption of TANFOR™ T and the shrinkage temperature of the leather

4.4.1.2 Tanning with organic tanning agents

In order to avoid chromium completely in leather processing, some organic tanning agents such as aldehyde tanning agents, tetrakis hydroxymethyl phosphonium salts, oxazolidine, vegetable tannin, dialdehyde polysaccharide, etc have also been investigated.

- Aldehyde tanning agents

In aldehyde tanning agents, glutaraldehyde and modified glutaraldehyde are commonly used in the tanning of pelt. Glutaraldehyde has a serious pungent odor and contains free formaldehyde, and the finished leather looks yellow. Modified glutaraldehyde has the same drawbacks as unmodified ones except for the improved color of leather. Schill & Seilacher Co. Ltd develops the pretanning system including Derugan 2000, Derugan 2020. Derugan 2000 is a modified glutaraldehyde with a certain self-buffering performance which can reduce the affinity of collagen and aldehyde group and promote itself to penetrate the pelt. Then Derugan 2000 reacts with collagen along with the rise of the pH of the float gradually. Derugan 2020 is a modified glutaraldehyde which can lenitively react with collagen and homogeneously permeate into the thicker pelt. The leather tanned by Derugan is fine, white and has good hydrothermal stability. Therefore, Derugan is used with the preferred vegetable tannin, syntans and fat-liquoring agents to meet the needs of car seat leather in the production of car seat leather. Moreover, G. Wolf, et al also studied the technology of the production of wet white based on glutaraldehyde, modified glutaraldehyde of BASF Co. Ltd which can markedly improve the quality of the finished leather.

- Tetrakis hydroxymethyl phosphonium salts

Tetrakis hydroxymethyl phosphonium salts (THP salts) are widely employed in the textile industries or other industries as fire retardants or bactericides and it was found to have tanning properties in the 1950s. Frequently-used Tetrakis hydroxymethyl phosphonium salts consist of tetrakis hydroxymethyl phosphonium sulfate (THPS), tetrakis hydroxymethyl phosphonium chloride (THPC), etc whose chemical structure is shown in Figure 4. 8. In recent years, it has been introduced as a tanning agent into leather industry. Results of research have shown phosphonium compounds have advantages such as good tanning properties (fast-penetration, almost complete exhaustion and achievement of a shrinkage temperature of 80℃), low toxicity, and high biodegradability. The mechanism of phosphonium compounds tanning is to form cross-linking bonds between the P-hydroxymethyl group and amino group of collagen so as to improve the hydrothermal stability of the leather. But the results of the examinations show phosphonium compounds would release formaldehyde during tanning, leading to an excessive amount of free formaldehyde in finished leather.

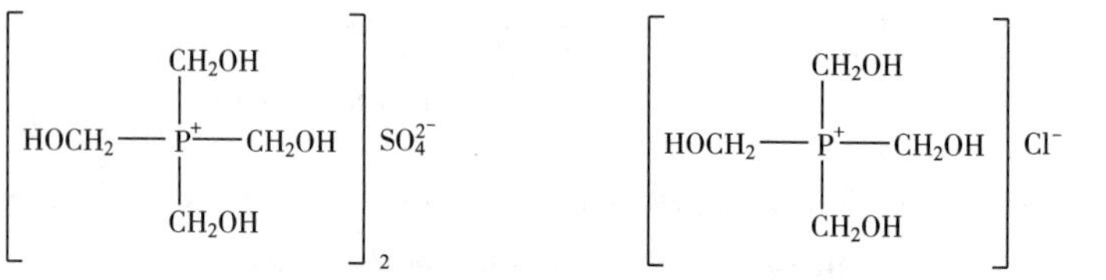

$$\left[HOCH_2 - \overset{\displaystyle CH_2OH}{\underset{\displaystyle CH_2OH}{P^+}} - CH_2OH \right]_2 SO_4^{2-} \qquad \left[HOCH_2 - \overset{\displaystyle CH_2OH}{\underset{\displaystyle CH_2OH}{P^+}} - CH_2OH \right] Cl^-$$

(a) Tetrakis hydroxymethyl phosphonium sulfate (b) Tetrakis hydroxymethyl phosphonium chloride

Figure 4. 8 The chemical structure of tetrakis hydroxymethyl phosphonium sulfate and tetrakis hydroxymethyl phosphonium chloride

- Oxazolidine

Oxazolidineis regarded as one of the best tanning agents in the 20th century. The shrinkage temperature of the leather tanned by oxazolidine can be 80-85℃ and the leather is white, full. Based on the formulation of synthetization and the structure of oxazolidine, it can also release formaldehyde during the tanning or the placing of the leather. The problem of free formaldehyde hasn't still solved over the years. In addition, the price of this type of tanning agent is too high to spread for application. Another important reason that restricts the development of wet white technology is that the tanning of organic tanning agent increases the electronegativity of wet white, which will hinder the uptake of anionic chemicals, like dyestuffs and fatliquoring agents, during post tanning processes and lead to poor quality of finished leather.

- TWT

In addition, Sichuan University and Sichuan Tingjiang New Materials Co. Ltd developed an amphoteric polymer tanning TWT which is used for the tanning of the pelt and has good tanning effect. The shrinkage temperature of the leather tanned by TWT can be between 80℃ and 85℃ so that it has good mechanical properties such as splitting, shaving, etc. Results of applied experiments show that it can directly tan the bated pelt without pickling. Thus it can shorten the process of leather-making and reduce the content of chloride ion in the effluent. Simultaneously the leather tanned by TWT has a stronger capacity of absorption and fixation for anionic materials such as re-tanning agents, dyestuff, fat-liquoring agents, etc because of its amphoteric characteristics. The tanning technology of TWT is shown in Table 4. 26.

Table 4. 26 Tanning technology of TWT based on sheepskin, goatskin or bovine hide

Process	Chemicals	Offer/%	T/℃	t/min	pH	Remarks
Pretreatment	Water	50	18-22		7. 0	
	Fat-liquoring agent	0. 5		20		
Tanning	TWT	4-5		120-240		Sheepskin or goatskin, 4%, 120min. Bovine hide, 5%, 240min
Basification	Sodium formate	1		30		
	Sodium bicarbonate	1-1. 2		3×15+30	7. 0-7. 5	
	Water (50℃)	100	40	180		
Stopping drum overnight, horse up						

- Vegetable tannins

As previously mentioned, vegetable tannins are considered to be eco-friendly owing to their natural origin and easy bio-degradation under the reaction of aspergillus niger. The shrinkage temperature of the leather tanned by the hydrolysable tannins can obtain 75-80℃. Whereas the shrinkage temperature of leather tanned with condensed tannin is between 80℃ and 85℃. Besides the color of the leather tanned by TARA is close to white, the leathers tanned by other vegetable tannins have a certain color such as light brown, dark brown, yellow brown, etc. Because vegetable tannin has a complicated structure and big molecular weight, the leather tanned by vegetable tannin is tight, stiff, which is good for the production of the moulding leather and is not suitable for the production of soft or Nappa leather.

- GTANOFIN F-90

In 2010, Clariant Chemcials (China) Ltd. developed and produced a tanning agent GTANOFIN F-90 which is a polymer synthesized by organic amine, fatty aldehyde with active hydrogen, allylic acid, allyl monomer with cation, etc. The structure of GTANOFIN F-90 is shown in Figure 4. 9.

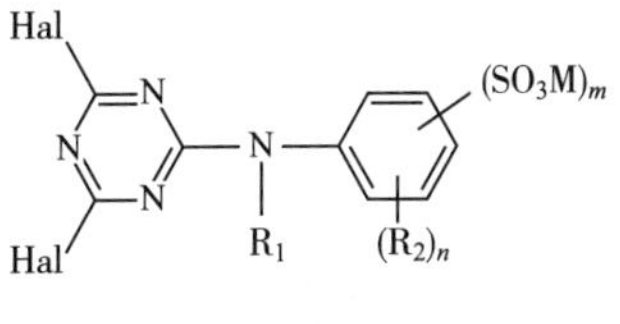

(Among the structure, Hal is Cl; R_1 is H, alkyl, alkoxyl group of C_{1-8}, or $\leftarrow C_{2-3}$Alkenylene—O—$\rightarrow_q$; R_2 is alkyl, alkoxyl group of C_{1-4}; m is 1or 2; n is 0 or 1; q is 1-10; M is H, alkali metal ion or NH_4^+.)

Figure 4. 9 The structure of GTANOFIN F-90

Liang Wenhua, et al studied the application of GTANOFIN F-90 in the tanning process of sheepskin and bovine hide by optimizing the tanning technology and characterizing it with a variety of indicators. Results figured that the tanning process was simple, the yield of leather was higher than that of chrome tanning leather, the shrinkage temperature of leather was between 75℃ and 80℃, and the white leather could be well retanned with different retanning agents. It was shown that the prospects of the chromium-free tanning process would be bright with GRANOFIN F-90.

- Dialdehyde polysaccharide

Polysaccharide is modified by chemical methods and is prepared into an organic tanning agent which can be used in the pretanning process of pelt or the retanning of the leather. Compared with the syntan based on the fossil resources, dialdehyde polysaccharide has remarkable advantages in terms of sustainable development. By the oxidation of periodate, Polysaccharide with vicinal diols is converted to biodegradable dialdehyde polysaccharide which shows the tanning properties by reacting with the amino groups of the collagen fibers to form covalent bonds. The shrinkage temperature of the leather tanned by dialdehyde polysaccharide can be 80℃ or so because of its high molecular weight and concentrated distribution.

In recent years, Shi Bi, et al made a profound study of the relation between the structure of dialdehyde polysaccharide (Figure 4.10) and its tanning properties and prepared the novel dialdehyde polysaccharide with multicomponent and broad distribution (Figure 4.11). In the tanning process of the novel dialdehyde polysaccharide, the homogeneity and degree of tanning are significantly improved so that the shrinkage temperature of the leather can obtain 85℃ and 90℃ and the leather has good softness, extensibility. The novel dialdehyde polysaccharide can be added into the drum with pickled pelt and tan the pelt. The tanning formulation of the novel dialdehyde polysaccharide is shown in Table 4.27.

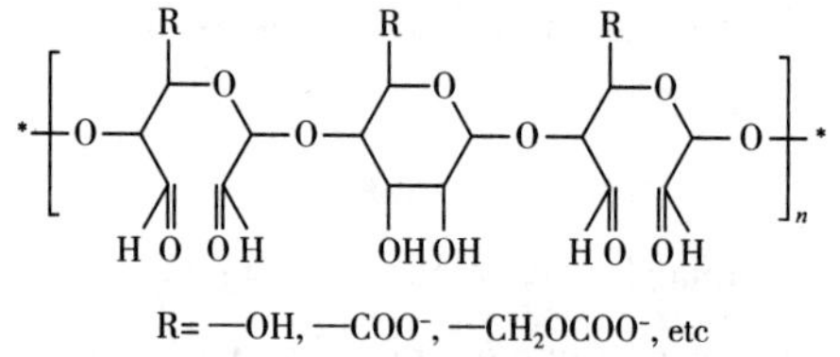

R= —OH, —COO^-, —CH_2OCOO^-, etc

Figure 4.10 The structure of dialdehyde polysaccharide

N=CH
N=CH
DAP
N=CH
N=CH

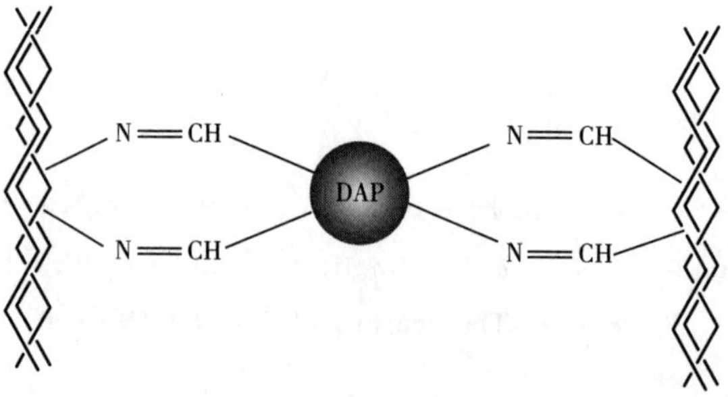

Figure 4.11 The cross-linking principle between dialdehyde polysaccharide and collagen

Table 4.27 Tanning technology of the novel dialdehyde polysaccharide based on bovine hide

Process	Chemicals	Offer/%	T/℃	t/min	pH	Remarks
Tanning	Pickling liquid	100			18-22	Based on twice the weight of pickled pelt
	Dialdehyde polysaccharide	4		240		
Basification	Sodium bicarbonate	1-1.2		3×15+30	8.0	
	Water (50℃)	100	40	240		
Stopping drum overnight, horse up						

However, the current wet white technologies face some bottlenecks such as weak tanning property and relatively strong electronegativity of wet white, resulting in a gap of quality between leather made by wet white and conventional chrome-tanned leather. Therefore each of other tanning agents has associated disadvantages. No one individual tanning agent has been able to match the properties of chromium.

4.4.2 Combination tannage of two chrome-free tanning agents

In order to further improve the hydrothermal stability of the chrome-free tanned leather and reach or is close to the quality of the chrome tanned leather, two or multi chrome-free tanning agents are matched to tan the pelt, realize the combination tanning and form the synergism of two or more tanning agents. Results of the investigations show that combination tannages are considered as a suitable alternative for chrome-free tanning system. These combination tannages include the combination tanning of different chrome-free metal tanning agents, different organic tanning agents or chrome-free metal tanning agents and organic tanning agents.

4-9

4.4.2.1 Combination tanning of different chrome-free metal tanning agents

B. Madhan, et al studied the tanning properties of the aluminium-zinc complex tanning agent which is prepared by aluminium sulfate, zinc sulfate, sodium citrate, sodium tartrate, triethylenediamine diglycolate, phthalic acid, etc. Results show that the shrinkage temperature of the leather tanned by aluminium-zinc complex tanning agent can obtain 90℃ and the physical, mechanical properties of the leather are close to that of chrome tanning leather. The sort of the masking agents can affect the tanning effect of the aluminium-zinc complex tanning agent. When hexamethylene tetramine acts as the masking agents, the shrinkage temperature of the leather tanned by aluminium-zinc complex tanning agent can obtain 95℃ and the quality of the leather is close to that of chrome tanning leather in terms of the physical, mechanical properties and sensory characteristic. Hexamethylene tetramine can improve the precipitation point of the aluminium-zinc complex tanning agent so as to be good for its permeation and combination in the pelt.

4.4.2.2 Combination tanning of different organic tanning agents

In order to replace the chrome tannage, many researchers investigate the tanning effect of combination tannage of different organic tanning agents. Combination tannage between vegetable tannin and aldehyde is one of the most important combination tannges. The leather tanned by combination tannages between vegetable tannin and aldehyde has some advantages such as high shrinkage temperature, low cost, biodegradation of leather shavings, etc. Aldehyde tannning agents include formaldehyde, modified glutaraldehyde, oxazolidine, etc. The leather tanned by combination tannages between vegetable tannin and formaldehyde is tight, low tear strength, brittle, because formaldehyde makes the connections among the collagen fibers hard to poor slip by cross-linking of methylene groups. Therefore, the aldehyde with aliphatic chains are used in the combination tannage.

- Combination tannage between vegetable tannin and modified glutaraldehyde

Shi Bi, et al studied the technology of the production of goatskin garment leather by adopting the combination tannage between vegetable tannin and modified glutaraldehyde and obtain the finished leather with high shrinkage temperature, fullness and softness. The technology not only eliminates the pollution of chromium (Ⅲ), but also partly overcomes the shortcomings of the vegetable tannin tanned leather retanned by formaldehyde such as poor mechanical performance, not resistant to storage, etc. They further investigate the combination tannage and its performance of the finished leather. The results are shown in Table 4.28. It can be seen from Table 4.28 that the optimal condition of combination tannage is firstly pre-treatment with syntan DS, then tanned by vegetable tannin and re-tanned using modified glutaraldehyde. In terms of the shrinkage temperature, tensile strength, tear strength of the leather, the combination tannage of vegetable tannin and modified glutaraldehyde is better than combination tannage of modified glutaraldehyde and vegetable tannin.

Table 4.28 Comparison of the properties of the finished leather tanned by different tannages

Tannage	T_s/℃	Tensile strength/ (N/mm^2)	Tear strength/ (N/mm)	Elongation under load/%	Water-absorbing quality in 15 minutes/%
Combination tannage of Veg-MG	95	15.8	21.0	38	131.4
Combination tannage of MG-Veg	91	11.3	16.4	42	153.5
Conventional chrome tannage	≥100	17.1	27.7	55	109.5
Standards of leather industry	≥90	≥6.5	≥18	25-60	—

- Combination tannage between vegetable tannin and oxazolidine

As previously mentioned, oxazolidine has good tanning properties. Because the function groups of oxazolidine can make the cross-linking reaction with the collagen and vegetable tannin and drastically improve the shrinkage temperature of the collagen, many researchers carried out a lot of research on the combination tannage between vegetable tannin and oxazolidine in recent years. Shi Bi, et al systematically studied the combination tannage between vegetable tannin and oxazolidine. Results show that combination tannage of vegetable tannin and oxazolidine is better than combination tannage of oxazolidine and vegetable tannin, condensed vegetable tannin is superior to hydrolysable vegetable tannin. The leather tanned by combination tannage between mimosa and oxazolidine has high shrinkage temperature (T_s>110℃) and boiling water resistance. When the consumption of mimosa and oxazolidine is 15%-20%, 6% respectively, the shrinkage temperature of the leather can exceed 110℃. Moreover, the shrinkage temperature of the leather can obtain more than 95℃ when the consumption of mimosa is 5%-10%. Thus the tannage is suitable for the production of shoe upper leather, garment leather, etc.

Shi Bi, et al studied the mechanism of reaction among vegetable tannin, aldehyde and collagen and have proved that the reaction between vegetable tannin and collagen is the synergistic effect of hydrogen bond and hydrophobic bond and physisorption of tannin colloid. The synergistic effect is the main factor of improving the hydrothermal stability of the leather. They further prove that aldehyde can react with 6, 8-bit of ring A in the molecules of condensed vegetable tannin and form covalent bonds. They has deeply explored the mechanism of reaction among vegetable tannin, aldehyde and collagen by the thermal analysis technique. By the comprehensive analysis of results of many experiments, the binding among vegetable tannin, aldehyde and collagen can be shown in Figure 4. 12 in the combination tannage between vegetable tannin and aldehyde. The basis of the combination tannage is the hydrogen bond between condensed vegetable tannin and collagen, covalent bond between aldehyde and the amino groups of collagen. Simultaneously condensed vegetable tannin takes part in the cross-linking reaction so as to form synergistic effect.

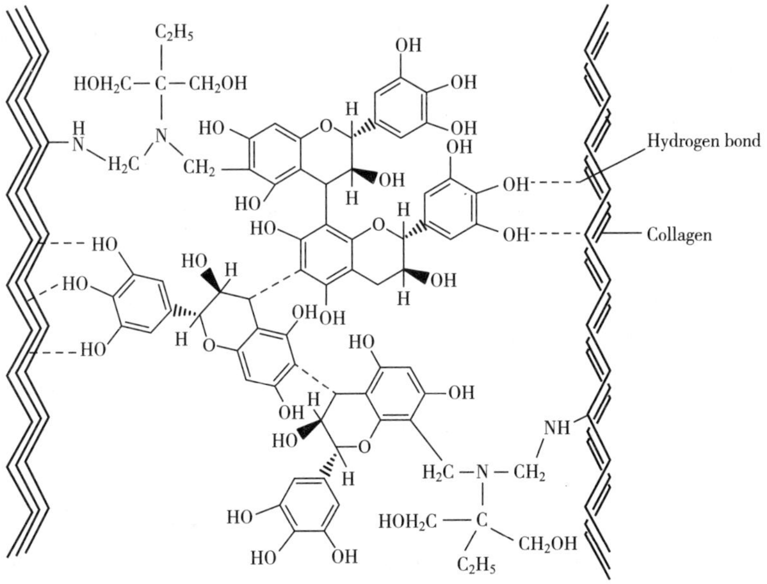

Figure 4. 12 Cross-linking mode of vegetable tannins-aldehyde combination tannage

Therefore, in the process of the combination tannage between vegetable tannin and aldehyde, condensed vegetable tannin permeated into the pelt forms multi-point binding of hydrogen bond with collagen. Then aldehyde reacts with the amino groups of collagen and forms Schiff's base and further occurs nucleophilic reaction with ring A of condensed vegetable tannin and forms stable cross-linking bonds.

The technology of the production of bovine chrome-free carseat leather has been found based on the combination tannage between vegetable tannin and aldehyde. The results of industrial tests show that the physical and mechanical properties of the leather can achieve the national standard and meet the needs of clients and be comparable with that of the chrome tanned leather. The technology eliminates the pollution of chromium (Ⅲ) in the effluent, sludge and leather goods, and reduces the discharge of salt by 40%-50% and saves water by 30%.

However, because the price of oxazolidine is high, the combination tannage of vegetable tannin and oxazolidine is not suitable for further promotion and application in the production of leather.

Combination tannage between Tetrakis hydroxymethyl phosphonium sulfate (THPS) and aldehyde or tannic acid is also one of the most important combination tannges. N. N. Fathima, et al studied the combination tannage between THPS and tannic acid, acet aldehyde or glutaraldehyde and found the shrinkage temperature of the leather can obtain more than 85℃ when the amount of THPS is 1.5% based on the weight of pickled pelt. The physical and mechanical properties of the leather are equal to that of chrome-tanned leather. Moreover, THPS can confer light color, high mechanical strength, good fastness to light and non-yellowing resistance for the leather.

4.4.2.3 Combination tannage between vegetable tannin and chrome-free metal tanning agent

As previously mentioned, the tanning effect of a single organic tanning agent or chrome-free metal tanning agent can't meet the need of the finished leather. Many researchers studied the combination tannages of organic tanning agents and chrome-free metal tanning agent such as the combination tannage between vegetable tannin and chrome-free metal tanning agents, the combination tannage between synthetic resin and chrome-free metal tanning agents, the combination tannage between oxidized polysaccharide and chrome-free metal tanning agents, etc.

In the combination tannages between vegetable tannin and chrome-free metal tanning agent, aluminium salt, zirconium salt, titanium salt and rare earth, etc are used and studied successively by researchers at home and abroad.

- Combination tannage between vegetable tannin and aluminium salt

He Xianqi, Covington, Kallenberger, et al separately studied the combination tannage between vegetable tannin and aluminium salt and its mechanism (Figure 4.13). They proposed that vegetable tannin can firstly bind with collagen by hydrogen bond and hydrophobic bond, then the succeeding metal ion can bind with the carboxyl group of the side chain of collagen and vegetable tannin by coordination bond so as to increase the effective cross-linking among collagen fibers and improve the hydrothermal stability of the collagen fibers. Covington, et al proposed that Matrix theory can contribute to high hydrothermal stability of the collagen fibers.

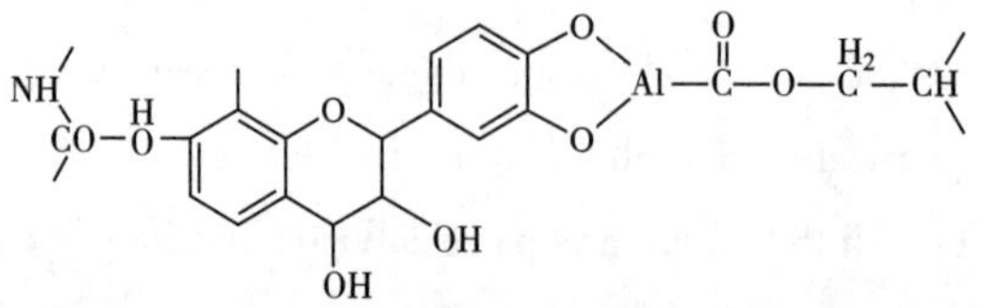

Figure 4.13 The reaction among vegetable tannins, aluminium and collagen in vegetable tannins and aluminium combination tannage

Polymer theory proposed by Hernandez J F, et al is also of valuable reference. The coordination interaction of metal ions causes vegetable tannin to penetrate into the pelt to form polymers which bind with the collagen fibers by mainly hydrogen bonds. The increasing of binding points and more hydrogen bonds can greatly improve the hydrothermal stability of the leather.

Combination tannage between vegetable tannin and aluminium salt can be processed by firstly pretanning of aluminium salt and then tanning of vegetable tannin or firstly pretanning of vegetable tannin and then tanning of aluminium salt. Tanning effects among different vegetable tannins with aluminium salt are shown in Table 4. 29. It can be seen from Table 4. 29, the shrinkage temperature of the leather of firstly pretanning of vegetable tannin and then tanning of aluminium salt is higher than that of firstly pretanning of aluminium salt and then tanning of vegetable tannin.

Table 4. 29 The shrinkage temperature of the leather by Combination tannage of Veg-Al or Al-Veg

Tannage	The shrinkage temperature of the leather/℃	Tannage	The shrinkage temperature of the leather/℃
Myrica extract-Aluminium	111	Aluminium-Myrica extract	90
Larch Bark Extract-Aluminium	95	Aluminium-Larch Bark Extract	87
Emblic extract-Aluminium	113	Aluminium-Emblic extract	96
Beefwood extract-Aluminium	107	Aluminium-Beefwood extract	93

For the tanning effect of combination tannage between vegetable tannin and aluminium salt, hydrolysable tannin such as TARA, Valonea extract, etc is better than condensed tannin. The quality of the leather tanned by combination tannage between vegetable tannin and aluminium salt is close to that of the chrome tanning leather when the amount of vegetable tannin is low. Vegetable tannin should have mild astringency, good permeability, appropriate color, and its consumption is 15% or so based on the weight of the limed pelt. Among many vegetable tannins, mimosa extract, TARA, etc is preferentially suitable for the combination tannage because they have mild astringency, rapid permeability, light color and strong capability of coordination with aluminium salt. The shrinkage temperature of the tanned leather can be more than 100℃.

The formulation of combination tannage between vegetable tannin and aluminium salt used in the production of bovine shoe upper leather is shown in Table 4. 30. In order to promote the permeation of vegetable tannin in the pelt, the pelt is firstly pretreated by glutaraldehyde or syntan before it is tanned by vegetable tannin. S. Vitolo, et al produce bovine shoe upper leather with high quality by adopting the pretreatment of glutaraldehyde or syntan, the combination tannage between TARA and aluminium salt and its formulation is shown in Table 4. 31. The shrinkage temperature of the finished leather tanned by combination tannage between TARA and aluminium salt exceeds 100℃. Their physical-mechanical properties are shown in Table 4. 32.

Table 4.30 The formulation of combination tannage between vegetable tannin and aluminium salt used in the production of bovine shoe upper leather

Process	Chemicals	Offer/%	T/℃	t/min	pH	Remarks
Pickle	Water	30	18-22			
	Sodium chloride	6				
	Sodium formate	1		5		
	Sulfuric acid	1		60	3.8	Sulfuric acid is diluted and cooled
Pretreatment	Anhydrous Sodium Sulfate	10		120	4.2	
Tanning	Assisted syntan orsulfited fish oil	2		30		
	Mimosa	15		180-240		
	Water	50	18-22	120	4.2	Drain
Washing, wringing, shaving						
Rewetting	Water	150	18-22			Based on the shaved leather
	Oxalic acid or EDTA	0.3		20		Drain and washing
Adjusting pH	Water	100	18-22			
	Formic acid	0.5		30	3.0	Drain
Tanning	Water	70	30			
	Anhydrous aluminum sulfate	10		60		
	Sodium acetate	1		30		
Basification	Sodium bicarbonate	1-1.5		150	3.8	Washing and horse up for 24 hours
Neutralization	Water	150	30			
	Sodium formate	1				
	Sodium bicarbonate	0.5		60	4.5	
The subsequent retanning, dyeing and fat-liquoring process is in accordance with conventional methods. The final pH of the float shouldn't beless than 4.						

Table 4.31 The formulation of combination tannage between TARA and aluminium salt used in the production of bovine shoe upper leather

Process	Chemicals	Offer/%	T/℃	t/min	pH	Remarks
Pickle	Water	50	18-22			
	Sodium chloride	4		5		
	Formic acid	1-1.5		60	2.8-3.0	Sulfuric acid is diluted (1 : 10)

Continued Table

Process	Chemicals	Offer/%	T/℃	t/min	pH	Remarks
Pretreatment	Glutaraldehyde	1		120	4. 2	
	Sodium bicarbonate	0. 5–0. 8			3. 8–4. 0	Drain
Washing	Water	200	25	10		Drain
Tanning	Water	50	18~22			
	TARA	20		60		Or the pretreated by 12% syntan for 150min
	Formic acid	0. 3–0. 5			3. 2	Drain
Horse up for more than 36 hours. Splitting, shaving						
Washing	Water	200	25	10		Drain
Tanning	Water	50	20			
	$Al_2(SO_4)_3 \cdot 4H_2O$	17				
	Citric acid	5. 8		120		
	Sodium bicarbonate	0. 8–1. 0		60	4. 0	Drain
The subsequent retanning, dyeing and fat-liquoring process is in accordance with conventional methods.						

Table 4. 32 The physical-mechanical properties of the finished leather tanned by combination tannage between TARA and aluminium salt under the condition of pretanning with different tanning agents

Item	Pretanned by glutaraldehyde	Pretanned by syntan	Standards of leather industry
Tension strength of the leather/(N/mm^2)	21. 3	27. 5	≥10
Load of tearing/N	80	52	≥50
Height of grain elongation of the leather	10. 5	11. 6	≥7

In terms of the mechanical properties and sensory index of the leather, the technologies are suitable for the production of high-quality shoe upper leather. The leather pretreated by glutaraldehyde and tanned by combination tannage between TARA and aluminium salt is softer, and fuller than the leather pretreated by syntan and tanned by the same methods. But the leather has coarse grain, stiff handle, poor extensibility, no resistance to being stored, and strong sensation of vegetable tannin tanned leather.

To solve the problems of the combination tannage between vegetable tannin and non-chrome metal salt, modifying vegetable tannin, pretanning of glutaraldehyde, syntan or modified starch or studying complex tannage based on the combination tannage are effective methods. Shi Bi studied the oxidation and degradation of valonia extract and obtained modified valonia tannin with certain molecular weight and filling properties. Modified valonia tannin is easy to penetrate into pelt and can save the tanning time. The leather tanned by combination tannage between modified valonia tannin and aluminium salt has high shrinkage temperature, fine and tight grain, soft and full handle, as well as has good formability. Therefore, the combination tannage between modified valonia tannin and aluminium salt has good application prospects. In addition, in order to solve the problems of dark color, slow permeability, etc of the leather tanned by vegetable tannin, Fathima, et al studied and adopted the combination tannage among gallic acid, aluminium salt and silicon. The shrinkage temperature of the leather can obtain 95℃. The physical-mechanical properties of the leather meet the needs of leather industry and the finished leather is softer, and smoother than those of the leather tanned by traditional chrome tannage.

- Combination tannage between Tetrakis hydroxymethyl phosphonium sulfate (THPS) and chrome-free metal salt

Combination tannage between Tetrakis hydroxymethyl phosphonium sulfate (THPS) and non-chrome metal salt is also one of the most important combination tannges. N. N. Fathima, et al studied the combination tannage between THPS and iron salt, aluminium salt, zirconium salt. The formulation of the combination tannage between THPS and zirconium salt is shown in Table 4. 33. They found the shrinkage temperature of the leather can obtain more than 85℃ when the amount of THPS is 1. 5% based on the weight of pickled pelt. The physical and mechanical properties of the leather are equal to that of chrome-tanned leather. Moreover, THPS can confer light color, high mechanical strength, good fastness to light and non-yellowing resistance for the leather.

Table 4. 33 The combination tannage between THPS and zirconium salt based on the pickled hide

Process	Chemicals	Offer/%	T/℃	t/min	pH	Remarks
Tanning	Pickling liquid	50	18-22		2. 8	
	THPS	1. 5		45		
	Zirconium oxychloride	10		60		
	Sodium tartrate	2. 5		30		
	Water	50		10		
Basification	Sodium formate	0. 5		30		
	Sodium bicarbonate	1. 0-1. 2		3×10+120	3. 8-4. 0	Dilute by 1 : 10
Stopping drum overnight, horse up						

- Combination tannage between nano-SiO_2 and oxazolidine, or other tanning agents

Fan Haojun, et al develop a tannage based on nano-SiO_2. Firstly, the nano-precursor is prepared by mixing tetraethoxysilane and a dispersing carrier including the polymer or modified oil. Then the nano-precursor is introduced into the gaps of the collagen fibers by the osmosis and diffusion of the dispersing carrier. At last, the nano-precursor is hydrolyzed and in-situ generate nano-SiO_2 to form the organic-inorganic hybridization with collagen by decreasing the pH of the float. The hybridization can improve the hydrothermal stability of the leather which includes the bonding reaction between nano-SiO_2 and the side group, peptide chain of histidine, tryptophan and the condensation reaction between the Si—OH group originated from the hydrolysis of the nano-precursor and the hydrogen group of side chain of the collagen. The tanning formulation of nano-SiO_2 is shown in Table 4.34. Results show that the shrinkage temperature of the leather tanned by 0.3% nano-SiO_2 based on the weight of bated pelt.

Table 4.34 The tanning formulation of nano-SiO_2 based on the bated hide

Process	Chemicals	Offer/%	T/℃	tmin	pH	Remarks
Pretanning	Water	50	18-22			
	Oxazolidine	2		120		
Horse up, wringing, splitting						
Tanning	Nano-SiO_2	0.3		60		
Acidification	Water	50		10		
	Formic acid	1.5		120-180	3.0-3.5	Dilute by 1:10
Stopping drum overnight, horse up						

Fan Haojun, et al also studied the combination tannage between nano-SiO_2 and other tanning agents based on the tanning characteristics of nano-SiO_2. They found that the high degree of pickling process can promote the penetration of nano-SiO_2 and increase the softness of the leather, combination tannage between nano-SiO_2 and THPS can effectively accelerate the penetration and absorption of the retanning agents, dyestuff, fat-liquoring agents, etc.

On the whole, the tanning technology of nano-SiO_2 accords with the needs of clean production of leather manufacture and has well developed and applied prospects.

- Combination tannage between vegetable tannin and rare earth

Shan Zhihua, et al studied the combination tannage between vegetable tannin and rare earth which can be processed by firstly pretanning of vegetable tannin and then tanning of rare earth. They can not only evenly permeate into the pelt, but also obtain good synergism and high shrinkage temperature. Rare earth has stronger binding capability with vegetable tannin than that of rare earth with collagen. The consumption of vegetable tannin and rare earth is based on the properties of the finished leather such as the shrinkage temperature of the leather, fullness, etc. In general, the optimum ratio of vegetable tannin and rare earth is 2 : 1. In order to increase the permeability of vegetable tannin, decrease the astringency of vegetable tannin and improve the precipitation pH of reaction between rare earth and vegetable tannin, vegetable tannin can be modified by moderately reducing the molecular weight or adding the auxiliaries of penetration, turbidity retarder, etc. The finished leather can meet the needs of leather industry.

- Combination tannage between vegetable tannin and other metal salt

Beside saluminium tanning agent and rare earth, Fe^{2+}, zinc, etc were used in the combination tannage. Shan Zhihua, et al explore the possibility of combination tannage between Fe^{2+} and phenol condensate. Results show that the tannage of firstly tanning of Fe^{2+} and then retanning of phenol condensate has strong synergism. The shrinkage temperature of the leather can obtain 85℃ and the leather has light color, good physical-mechanical properties. Saravanabhavan, et al tan the sheepskin pelt by adopting gallic acid, zinc salt and sodium silicate. The shrinkage temperature of the leather can obtain 85℃ and the physical mechanical properties of the leather are close to that of the chrome-tanned leather.

- Combination tannage between synthetic resin and aluminium sulfate or titanium sulfate

S. Gangopadhyay, et al synthetized four synthetic resins and studied their combination tannage with chrome-free metal tanning agents. Synthetic resin Ⅰ was synthetized by urea, phenol, phenol sulfonic acid and formaldehyde. Synthetic resin Ⅱ was synthetized by dicyandiamide, phenol, phenol sulfonic acid and formaldehyde. Synthetic resin Ⅲ was synthetize by urea, salicylic acid, sulfosalicylic acid and formaldehyde. The tanning technology of synthetic resin-chrome-free metal tanning agents based on the pickle is shown in Table 4. 35. The shrinkage temperature of the leather can obtain more than 100℃ which shows the synthetic resin can evidently improve the tanning effect. The function of Synthetic resin Ⅱ is better than that of Synthetic resin Ⅰ. In addition, the sensory characteristics and physical-mechanical properties of the leather can achieve the standard of chrome-tanned leather.

Table 4. 35 The tanning technology of synthetic resin-chrome-free metal tanning agents based on the pickle

Process	Chemicals	Offer/%	T/℃	t/min	pH	Remarks
Pretanning	Pickling liquid	50	18-22			
	synthetic resin	6		240		Stopping drum for over night
Washing	Water	150	18-22	20		Drain
Tanning	Water	50	18-22	20		
	Aluminium sulfate or titanium sulfate masked by citric acid	1. 7% (Based on oxide)		120-240		
Basification	Sodium bicarbonate	1-1. 2		3×15+30	4. 6-4. 7	
	Water (50℃)	100	40	180		
Stopping drum overnight, horse up						

- Combination tannage among oxidized polysaccharide and zirconium salt, aluminium salt

As far as the defect of zirconium-aluminium tanning agent in tanning properties is concerned, Yu yue, et al prepared an oxidized polysaccharide tanning agent with more hydrogen groups, aldehyde groups, carboxyl groups and appropriate molecular size based on natural polysaccharide deeply oxidized and degraded by hydrogen peroxide using copper - iron salt as a catalyst. Combination tannage among oxidized polysaccharide, zirconium salt and aluminium salt can promote the pelt to absorb zirconium salt, aluminium salt and improve the shrinkage temperature, the quality

4-10

of the finished leather because oxidized polysaccharide, zirconium salt and aluminium salt form multi-point cross-linked network structure among collagen fibers. Zirconium salt and aluminium salt can coordinate the amino groups, carboxyl groups and hydrogen groups of collagen fibers. Meanwhile, oxidized polysaccharide can react with collagen fibers by hydrogen bonds and electrovalent bonds. The cross-linking mode and process of the combination tannage is shown in Figure 4. 14.

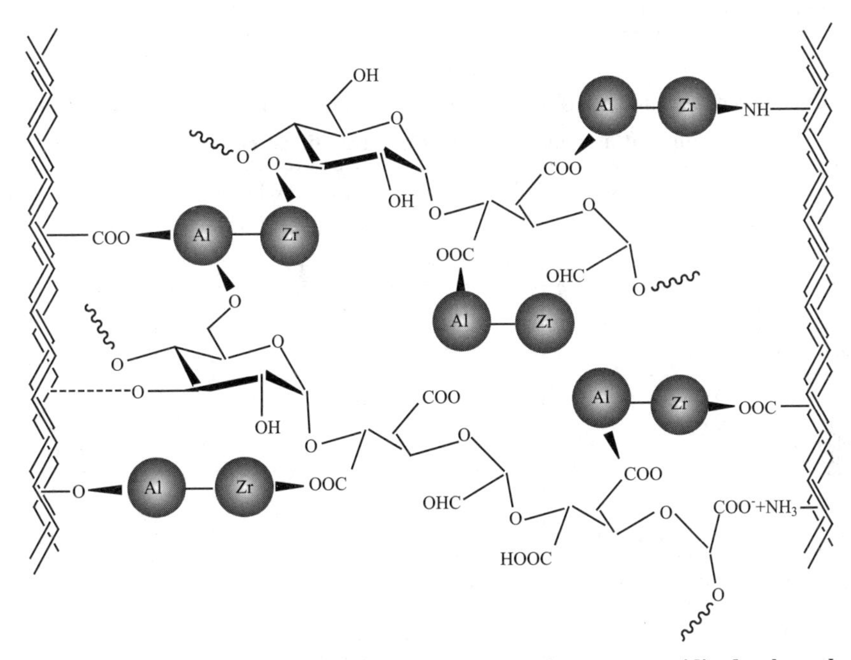

Figure 4. 14 The cross-linking mode of combination tanning among oxidized polysaccharide, zirconium salt and aluminium salt

Using the bovine hide pickle as materials, combination tanning among oxidized polysaccharide, zirconium salt and aluminium salt are shown in Table 4. 36. Comparing with the traditional chrome tannage, the relevant indexes of the process of the combination tannage are shown in Table 4. 37. It can be seen from Table 4. 37 that the absorption of tanning agent obtains 94. 4% which is good for decreasing the metal ion concentration in the effluent. The leather tanned by the combination tannage has a strong affinity for anionic retanning agents, dyestuff, fat-liquoring agents, etc and excellent mechanical properties and sensory performance so as to meet the needs of the production.

Table 4. 36 The tanning technology of combination tanning among oxidized polysaccharide, zirconium salt and aluminium salt on the bovine hide pickle

Process	Chemicals	Offer/%	*T*/℃	*t*/min	pH	Remarks
Tanning	Pickling liquid	50	18-22			
	Al-Zr tanning agent	2 (Based on oxide)				
	Oxidized polysaccharide	1. 6		240		

Continued Table

Process	Chemicals	Offer/%	T/℃	t/min	pH	Remarks
Basification	Magnesium oxide	0.9		3×30	~3.0	
	Sodium bicarbonate	2.5-3.0		4×15	3.8-4.0	
	Water	200	40	120		Drain
Stopping drum overnight, running 30 minutes, pH 4.0-4.2, horse up						

Table 4.37 Comparison of tanning effects between combination tannage among oxidized polysaccharide, zirconium salt and aluminium salt and traditional chrome tannage

Item	Combination tannage among oxidized polysaccharide, zirconium salt and aluminium salt	Traditional chrome tannage
Absorption of tanning agent/%	94.4±0.6	71.4±0.5
Content of metallic oxide in the leather/%	5.3±0.3	4.6±0.2
T_s of the leather/℃	86.0±0.8	113.3±0.6
Gain of the leather/%	97.8±0.4	95.6±0.9
pI of the leather	7.1	7.1
Absorption of the leather for the retanning agent/%	91.2±0.9	86.9±0.9
Absorption of the leather for the fat-liquoring agent/%	83.2±0.6	65.9±0.6
Softness of the leather/mm	7.7±0.4	7.6±0.5
Tension strength of the leather/(N/mm)	16.2±1.1	13.5±0.3
Tear strength of the leather/(N/mm)	44.5±0.9	39.7±2.2
Bursting strength of the leather/(N/mm)	247.6±17.5	180.7±14.8

In addtition, Guo Xueru, et al propose that oxidized polysaccharide ligands OP1 and OP2 with different carboxyl group contents and relative molecularweights were used to coordinate with aluminum salt and zirconium salt (AZ) for tanning. These two tanning technologies were compared with conventional citric acid (CA)-AZ tanning and chrome (Cr) tanning technologies in terms of the performance of tanned leather, the performance of crust leather and pollution loads. Results indicate that the relative molecular weight of the ligand has a greater influence on the tanning performance than that of carboxyl content. OP2 with M_w around 7,000 was used as the ligand of AZ. The tanned leather has a high shrinkage temperature (81.2℃), strong positive electricity (isoelectric point higher than 7.0), and good absorption for post-tanning chemicals (more than 80%), thereby resulting in excellent physical properties of OP2-AZ crust leather, which is second only to chrome tanned crust leather. OP1 (M_w around 1,000) and CA with lower molecular weight show poorer tanning performance than OP2 when they coordinate with AZ. In addition, the biodegradability of wastewater from OP2-AZ tanning technology is better than that of wastewater from CA-AZ and Cr. Therefore oxidized polysaccharide-aluminum-zirconium complex tanning technology is expected to become a practical and clean chrome-free tanning technology.

4.4.3 Complex combination tannage among more chrome-free tanning agents

4.4.3.1 Complex combination tannage with phosphonium compounds, vegetable tannins and aluminium tanning agent

Among the numberless combination tannages that are currently exploited, the vegetable tannins and oxazolidine combination tannages, the vegetable tannins and aluminium tanning agent combination tannages are one of the most promising options. They have different mechanisms for improving the stability of collagen. In its process, oxazolidine and aluminium act as cross-linkers so that it is possible to achieve a hydrothermal stability comparable to that of chrome-tanned leather. But based on the results of the semi-industrial scale, the shrinkage temperature of the leather tanned by combination tannage of veg-Al or veg-Oxa might be decreased by 20℃ after retanning, dyeing and fatliquoring, and this strongly affects the properties of the products. So the purpose of our research and paper is to develop a complex combination tannage that can reduce the effect of the post-tanning processes on decreasing shrinkage temperature, so that the tannage might be useful as a substitute for chrome tanning.

In order to further improve and protect high shrinkage temperature, good softness, fullness, handle, etc of the finished leather, Luo Jianxun, et al adopt the pretanning of phosphonium compounds and have investigated the use of phosphonium compounds in combination tannages with vegetable tannins and aluminium. The phosphonium compounds-veg-Al tannage was applied to hide powder and pickled goatskin. The shrinkage temperature of the leather obtained by complex combination tannage with phosphonium compounds, vegetable tannins and aluminium tanning agent is 110℃ which is much higher than that of combination tannages using vegetable tannins and aluminium tanning agents alone. The fiber bundles are better separated in ternary combination tannage than those in single tannage or binary combination tannage as judged from SEM images. Based on T_s, contraction in length, tensile strength and tear strength of the leather after heat-aging, a satisfactory ternary combination tannage sequence is confirmed as phosphonium compounds, vegetable tannins, and aluminium tanning agent. The optimized offers of phosphonium compounds, vegetable tannins, and aluminium tanning agents are 3.5%, 7% and 3.5 % based on pickled goatskin pelt. The complex tannage was conducted on an industrial scale and it gave a superior performance.

4.4.3.2 Complex combination tannage with vegetable tannins, aluminium tanning agent and oxazolidine

A new complex combination tannage based on combination tannages of veg-Al and veg-Oxa has been developed as a chrome-free tannage. The veg-Al-Oxa tannage was applied to hide powder and pickled goatskin. The shrinkage temperature of the finished leather obtained by veg-Al-Oxa complex combination tanning is 112℃, which is much higher than that of binary combination tannages. The fiber weave is better separated in ternary combination tannage than in a single tannage or binary combination tannage as judged from SEM images. Based on T_s, length contraction rate, tensile strength and tear strength of the leather on heat-aging, a good ternary combination tannage sequence is confirmed to be vegetable tannins, aluminium tanning agent and oxazolidine. The optimized offers of vegetable tannins, aluminium tanning agent and oxazolidine are 10%, 3% and 3% based on pickled goatskin pelt.

4.4.3.3 Complex combination tannage among multi-chrome-free metal salts

Based on the tanning characteristics of iron salt, zirconium salt, aluminum salt, titanium salt, etc, multi-chrome-free metal salts can be matched to form complex tanning agents or to directly be used in the tanning process in a certain order respectively.

Lan Yunju, et al developed and prepared LTA-1 chrome-free metal tanning agent and its auxiliaries based on zirconium tanning agent, titanium salt tanning agent and aluminum tanning agent. In the tanning process of LTA-1, the synergistic effect is created among three tanning agents. The quality of the finished leather is close to that of the chrome tanned leather. The technology has been used for the production of garment leather, shoe upper leather, furniture leather, etc.

Dan Weihua, et al also designed and developed the Zr-Al-Ti complex tanning agent based on zirconium sulfate, titanium sulfate, aluminum sulfate and sodium citrate. Then they systematically studied the tanning mechanism of Zr-Al-Ti metal complex and proposed a bimodal hypothesis based on the mode of self-loading of molecules by molecular regulation and the mode of Unit-Construction in the tanning system of chrome-free metal complex tanning agent and propose the reactions between the complex tanning agents and collagen include physical filling and chemical binding. Zr-Al-Ti metal complex can react with carboxyl group, hydrogen group, amino group, etc of the side chain of collagen fibers, as shown in Figure 4.15.

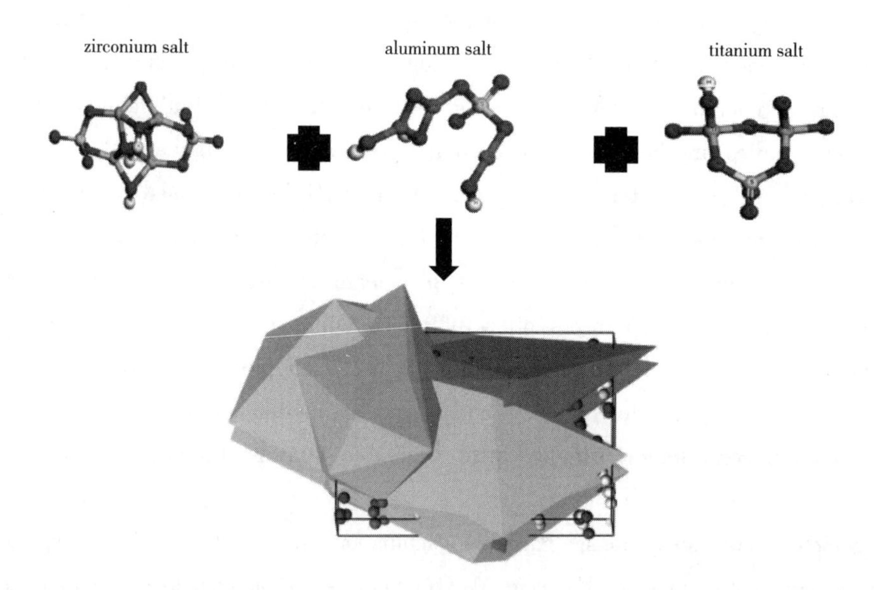

Figure 4.15 Forming process of Zr-Al-Ti metal complex

Wen Huitao, et al designed the pretanning, main tanning and retanning of cattle hide upper leather using pickled cattle skin and Fe-Zr-A1 complex tanning agent as the main raw material. Results showed that the pretanning with 1% glutaraldehyde GT50, the main tanning with 26% Fe-Zr-A1 complex tanning agent, and the retanning with 7% Fe-Zr-A1 complex tanning agent, 6% acrylic polymer tanning agent, 2% amino resin tanning agent and 5% vegetable tanning agent, the physical-mechanical properties and sensory properties of the finished leather were closed to those of the chrome tanned leather. The production process of Fe-Zr-Al complex tanned cattle hide upper leather was in line with the concept of ecological design, and it has very important realistic significance to improve the level of clean production in leather industry and promote the progress of leather engineering technology.

4.5 Clean technology of dyeing, retanning and fat-liquoring process

To further improve the performance of the tanned leather and achieve the function, and fashion of the finished leather, the process of retanning, dyeing and fat-liquoring, etc are necessary for the shaved wet blue or wet white. A lot of organic compounds such as retanning agents, dyestuff, fat-liquoring agents, etc are used during the operation. The degree of the absorption of the organic compounds directly affects the content of organic matter in the effluent. Therefore environment-friendly, Biodegradable materials and corresponding high absorption technologies are always one of the most important goals and strategic directions.

4-11

4.5.1 Dyestuff and its high absorption technology

As previously mentioned, dyeing process is one of the most important processes in leather manufacture which can improve the appearance, color, performance of the leather and meet the fashion style.

According to the needs of markets and the preservation of the environment, some azo dyestuff can't be used in the dyeing process which can be reduced to aromatic amine with carcinogenic effects on humans or animals under certain conditions. 24 kinds of aromatic amine with carcinogenic effects are shown in Table 4.38. They are prohibited in REACH regulations which have 210 related dyes. Therefore the dyestuff in the dyeing process should be in accord with the relevant laws or regulations.

In addition, how to improve the dye-uptake of dyestuff, eliminate or lower the effect of light color and reduce thecontent of residual dyestuff in the effluent is always the purpose of high absorption of dyestuff and one of projects during the production of leather manufacture. Now the dyeing auxiliaries, dye-fixatives, adjusting the pH of the float, the technology and new methods of dyeing, etc are adopted in order to achieve the high absorption of dyestuff.

J. Kanagaraj, et al developed NPP dyeing auxiliaries which are synthesized by an amino acid derivative and acrylate. The absorption of dyestuff can obtain 99.10% when theconsumption of NPP is 2% in the dyeing process. Dai Jinlan, et al synthesized the dyeing and fat-liquoring auxiliaries for high absorption. Results of application that they can achieve better absorption for the dyestuff and fat-liquoring agents and the color of the leather is dark.

Table 4.38 Aromatic amine with carcinogenic effect on human or animal

No.	Chemicals	CA	No.	Chemicals	CA
1	4-Aminodiphenyl	94-67-1	13	*p*-Kresidine	120-71-8
2	Benzidine	92-87-5	14	4,4'-Methylene-bis (2-choroaniline)	101-14-4
3	4-Choro-2-toluidine	95-69-2	15	*o*-Toluidine	95-53-4
4	2-Naphthylamine	91-59-8	16	2,4-Toluylenediamine	95-80-5
5	*o*-Aminoazotoluene	97-56-3	17	*p*-Chloroaniline	106-47-8
6	4,4'-Diaminodiphenylmethane	101-779	18	4,4'-Oxydianiline	101-80-4
7	2-Amino-4-nitrotoluene	99-55-8	19	4,4'-Thiodianiline	39-65-1
8	2,4-Diminoanisole	615-05-4	20	2,4,5-Trimethylaniline	137-17-7
9	3,3'-Dichlorobenzidine	91-94-1	21	*p*-Phenylazoaniline	60-50-3
10	3,3'-Dimethylbenzidine	119-93-7	22	*o*-Anisidine	90-04-00
11	3,3'-Dimethoxybenzidine	119-90-4	23	2,4-Xylidine	95-68-1
12	3,3'-Dimethyl-4,4 diaminodiphenyl-methane	838-88-0	24	2,6-Xylidine	87-62-7

Dye-fixatives are also important dyeing auxiliaries. Now frequently-used Dye-fixatives are mainly compounds with polycationic structure. Anionic groups of dyestuff can react with the cationic groups of dye-fixatives by ionic bond and form insoluble salt so as to improve the water resistance and wet fastness of dyestuff. In general, dye-fixatives are used in the later stage of dyeing process of the chrome-tanned leather, and the pH of the float is adjusted to 3.8 or so by formic acid solution. Z & S Co. Ltd develops and produces DR dye-fixatives which have strong positive charge and effectively, drastically improve the absorption of anionic dyestuff, retanning agents, fat-liquoring agents, etc. In addition, Bao Lihong, et al synthesized MAPA series dye-fixatives and studied their application in the dyeing process of chrome-tanned leather. Results show that the dye-uptake of dyestuff is improved from 85.2% to 97.4%, the penetrability of dyestuff, the grade of dry/wet-rubbing fastness of the finished leather is raised from 75.3% to 94.5% and from 4 to 5 respectively. Chen Hua, et al studied the application of DCA-1 dye-fixatives in the later stage of dyeing process of the chrome-tanned leather. Results show that DCA-1 can improve the grade of dry/wet-rubbing fastness of the finished leather, and reduce the usage of dyestuff and obtain the limpid, transparent effluent. Moreover, an aluminium tanning agent can be used as dye-fixatives to improve the absorption of dyestuff because of its strong positive charge.

In addition, the technology and new methods of dyeing are very important for the dyeing effect. In order to improve the absorption, penetrability for the dyestuff and increase the brilliance of color, dyeing process can be arranged before retanning and fat-liquoring process and then the dye-fixatives are used in the later stage of dyeing process. In order to further improve the absorption for the dyestuff and binding between the dyestuff and the leather, the new methods of dyeing are adopted by increasing the active points of the leather. For instance, the shaved wet blue is pretreated by active auxiliaries with double functional groups so as to increase positive charge of the leather and form the firm binding between dyestuff and collagen fibers. Besides the dyeing methods, other dyeing methods are studied or studying and applied in the dyeing process such as supercritical dyeing, ultrasonic dyeing, electro-chemical dyeing, through dyeing, etc. They can all improve the absorption, penetrability for the dyestuff to a certain extent.

4.5.2 Retanning agents and the high absorption technology

As previously mentioned, retanning is a very important process during leather manufacture, which is known as "the golden touch" in modern leather processing. The retanning process can not only improve the physical-mechanical properties and sensory performance such as softness, fullness, elasticity, tightness of grain, etc, but also improve the degree of the loose grain and reduce the location difference. The common retanning agents include mineral tanning agents, acrylic polymer retanning agents, syntans, amino resin retanning agents, vegetable tannin, protein retanning agents and other retanning agents. The ecological technology of retanning process is the preparation of the series of retanning agents with low salt, low free formaldehyde, low or no organic solvent, low or no free phenol, being easy to be degraded, improving the absorption, binding rate of retanning agents. Thus it can decrease the content of free formaldehyde, volatile organic compounds (VOC) of the leather, inorganic salts, COD, the total solid content, etc.

Therefore, developing ecological retanning agents is an important way to achieve ecological leather-making. Now the retanning agents based on biomass are studied and gradually applied in the production such as TRUPOTAN 05L, 08L (TRUMPLER), other polymer retanning agents based on hydrolytic collagen or oxidized starch, etc. The protein retanning agents not only have good retanning and filling properties, but also give the perfect handle, vapor permeability to the leather.

Simultaneously kaolin also acts as an important component in tanning agents or retanning agents which have good retanning and filling properties. In addition, syntans and amino resin retanning agents with low or no free formaldehyde/salt are also studied and gradually applied in the production of leather manufacture.

4.5.3 Fat-liquoring agents and their high absorption technology

Fat-liquoring process is one of the most important processes in leather-making which can give good softness, elasticity, fullness, physical-mechanical properties to the leather. The fat-liquoring agent is one of the materials with the largest consumption during the production of leather manufacture. Therefore, the absorption and binding between fat-liquoring agents and the leather are one of the most interesting aspects in the fat-liquoring process. If the absorption and binding of fat-liquoring agents are low, residual fat-liquoring agents can lead to the increase of COD, BOD in the effluent. Therefore, it is an important topic how to improve the absorption and binding of fat-liquoring agents and achieve the clean and ecological development of the fat-liquoring process. Developing the high absorptive, biodegradable fat-liquoring agents or auxiliaries and the fat-liquoring technology is the only way to achieve the clean and ecological development of the fat-liquoring process.

4.5.3.1 High absorptive and biodegradable fat-liquoring agents or auxiliaries

In order to meet the needs of the quality of the finished leather and environmental protection, the intrinsic fat-liquoring agents should be assessed and chosen, the structure of the new fat-liquoring agents should be designed and developed such as complex fat-liquoring agents, fat-liquoring agents with more functions, etc.

The intrinsic fat-liquoring agents include the modifiers of plant oil, animal oil or fat, mineral oil or their combinations and synthesized fat-liquoring agents, etc. They can give good softness, and physical-mechanical properties to the leather. Meanwhile, the different fat-liquoring agent has its own characteristics such as fullness, oily feel, etc. Therefore, the fat-liquoring agents are chosen and matched according to the state of shaved wet blue or wet white and the needs of the finished leather.

The main components of fat-liquoring agents are neutral oil (plant oil, animal oil or fat, mineral oil, etc), surfactant and other additions. In general, neutral oils are easy to be degraded because of their natural characteristics. Therefore, surfactant is the main component affecting the biodegradable properties of fat-liquoring agents. The whole process of bio-degradation includes primary degradation, secondary degradation, and final degradation. During the primary degradation, the parent structure and the characteristics of surfactant disappear or undergo changes. Then it is degradated to the substance without environmental pollution. Surfactants can be fully converted to inorganic substances such as water, carbon dioxide, ammonia, etc. The biodegradable properties of surfactants are closely related to their structure. Therefore the degree of biodegradable properties of surfactant can be inferred from its chemical structure.

The amino acid type, betaine type, amidopropyl betaine type zwitterionic surfactant, etc all have good biodegradability. In general, nonionic surfactants also have good biodegradable properties based on the length of chain of ethoxy group and linearity of alkyl chain. The degradation degree of alkylphenol ethoxylates is inferior to that of the surfactant with straight chains, non-aromatic hydrophobic group because the former has the structure of a branched alkyl and benzene ring. The degree of primary degradation of alkylphenol ethoxylates rises as the decreasing of the number of ethylene oxide addition (EO), increases the straightness and length of alkyl and the situation of phenolic group near alkyl group. The biodegradable properties of cationic surfactant are complicated because it has antibacterial activity and is easy to be adsorbed on the suspended solid. For the biodegradable properties of anionic surfactant, Carolyn, et al studied the biodegradable performances of more linear alkylbenzene sulfonates containing alkyl chains with different numbers of carbon atoms and corresponding replacement position of different benzene ring. Results show that the degradation rates of the linear alkylbenzene sulfonate are accelerated as the number of carbon atoms in alkyl chains of the linear alkylbenzene sulfonate rises when the substitution site of benzene rings in the alkyl chains is consistent. However, the degradation rates of the linear alkylbenzene sulfonate is fast as the substitution site of benzene ring in the alkyl chains is closer when the carbon atoms of alkyl chains is defined. On the whole, linear alkylbenzene sulfonates are easy to be biodegraded. When the alkyl chains of alkyl sulfonates are branched chain, their biodegradable performances is different from that of linear alkylbenzene sulfonates. According to the research results, the subsequence of the degree of biodegradable properties of anionic surfactants from high to low: Linear fat soaps, higher fatty alcohol sulfate, linear alcohol ether sulfate (AES), linear alkyl sulfonate or linear alkenyl sulfonate (AS, SAS, AOS), linear sodium alkylbenzene sulfonate (LAS), branched higher alcohol sulfate, branched ether sulfate, branched alkyl sulfonate (ABS).

Therefore the surfactants are chosen in the fat-liquoring agents according to the emulsifying and biodegradable properties.

4.5.3.2 High absorptive fat-liquoring technology

In order to improve the softness and fullness of the leather, developing new fat-liquoring agents and carrying out the fat-liquoring process by more steps are the only ways to ensure the distribution of fat-liquoring agents in the leather and improve the fat-liquoring effect. Besides the main fat-liquoring process, the fat-liquoring process by more steps can be implemented in the pickling, tanning, retanning, neutralizing, and dyeing process. In order to meet the needs of the fat-liquoring process by more steps, the fat-liquoring agents should have the characteristics of electrolyte resistance and theiremulsions are stable under the condition of acid, salt, chromium (Ⅲ), etc. Meanwhile, the match between the fat-liquoring agents and other chemical materials should be considered.

4.6 Clean technology of finishing process

4-12

Finishing is also an important process during leather manufacture. After the leather is finished, the resultant leather has uniform color, ideal gloss, comfortable handle, and excellent physical performance so as to meet the demands of the performance of the products and obtain excellent properties such as fastness to water, heat, low temperature, solvent, dry/wet rub, etc. Besides the needs of the physical performance of the finishing layer, some chemical indexes such as volatile organic compounds (VOC), chromiun (Ⅵ), plead (Ⅱ), Cd (Ⅱ), perfluorooctane sulfonate (PFOS), etc are also required and limited in recent years.

Ecological technologies of finishing process are mainly water-based finishing systems including green, environment-friendly finishing, water-based materials, formulations and its operation.

As previously mentioned, the current finishing materials have pigment or metal complex dyestuff, forming materials, water or solvent and auxiliaries. Yellow or red inorganic pigment or relevant metal complex dyestuff may cause the index of chromium (Ⅵ), plead (Ⅱ), Cd (Ⅱ), etc exceed the standard. Forming materials include acrylic emulsion, polyurethane emulsion, nitrocellulose emulsion, protein adhesive and compact resin. In the preparation of polyurethane emulsion, nitrocellulose emulsion, is firstly reacted in the organic solvent and then is emulsified by the way of internal or external emulsification. Therefore, the residual solvent is easy to cause the content of volatile organic compounds

(VOC) to exceed the prescribed limit. Organic solvent and some auxiliaries also cause the content of volatile organic compounds (VOC) to exceed the prescribed limit. Therefore, in the preparation of the finishing materials and the formulation of finishing process, the affection of organic solvent to the content of volatile organic compounds (VOC) in the leather should be taken into account.

In addition, the cross-linking agent is one of the most important auxiliaries which can react with the linear polymers to form the reticular structure so as to improve the physical performances of finishing layers. The common corss-linking agents have organometallic compound, aldehydes, epoxy compound, azoproridine, isocyanate, carbodiimide, etc. Sun Jing, etc studied the properties of these cross-linking agents. Results are shown in Table 4.39, and Table 4.40. It can be seen from the table, the different cross-liking agents show the relevant properties under the different conditions. On the whole, the developing tendency of the cross-linking agent is non-toxic, efficient, excellent, cheap and less amount and has special cross-linking effect. Therefore the developing novel efficient and non-toxic cross-linking agents will mainly be the direction during ecological technology of leather manufacture. Moreover, the solidification technology of the electron beam or ultraviolet rays (UV) can be studied and applied in the finishing process which can completely avoid the emission of volatile organic compounds (VOC). The finishing formulation includes photo-initiators, oligomer, monomer, dispersion and other auxiliaries. Photo-initiators can be derivatives of alkyl ketone, aryl ketone, benzoin, benzoin ether, monohydroxyalkyl acetone, and dipropyl ketone. The photo-initiator can produce radicals under the irradiation of ultraviolet light which makes radical polymerization between oligomer and monomer and further forms the cross-linking structure so as to obtain the function of forming films of the finishing agents. However, the principle of the solidification technology of electron beam is that oligomer can produce radicals under the irradiation of electron beam which makes radical polymerization between oligomer and monomer and further forms the cross-linking structure.

Table 4.39 The swelling rate of the film under the condition of different cross-linkers

Name of cross-linkers	Swelling rate/%		
	Water	Butyl acetate	Acetone
No cross-linkers	56	66	45
Azoproridine	8	25	0
Epoxy compound	20	50	1
Isocyanate	35	54	8
Isocyanate+Carbodiimide	19	42	0
Carbodiimide	13	40	0

Table 4.40 Relation between the amount of different cross-linkers and their rubbing fastness

Item	Azoproridine		Isocyanate		Carbodiimide		Epoxy compound	
	amount/%	Grade	amount/%	Grade	amount/%	Grade	amount/%	Grade
Fastness to dry rubbing	2.0	3.5	10.0	4.0	10.0	3.5	10.0	3.0

Continued Table

Item	Azoproridine		Isocyanate		Carbodiimide		Epoxy compound	
	amount/%	Grade	amount/%	Grade	amount/%	Grade	amount/%	Grade
Fastness to wet rubbing	2.0	2.5	10.0	3.0	10.0	2.5	10.0	1.5
Fastness to perspiration	2.0	2.5	10.0	2.0	10.0	2.0	10.0	1.5
Fastness to gasoline rubbing	2.0	1.5	10.0	2.5	10.0	1.0	10.0	1.0

As far as the cost of the solidification technology of the electron beam or ultraviolet rays is concerned, the cost of the solidification technology of ultraviolet rays is lower than that of the electron beam. At the same time, the solidification technology of ultraviolet rays has some advantages such as fast speed, saving energy, environmentally friendly, etc.

The finishing operations are one of the most important ways of ensuring the finishing effect. As previously mentioned, the finishing operations include padding, brushing, spraying, rolling, curtain finishing, etc. The characteristics of these operations are shown in Table 4.41.

Table 4.41 The characteristics of padding, spraying, rollering, curtain finishing operations

Methods	Put-on/ (g/sqft)	Wastage	Control of QTY put-on	Degree of evenness	Efficiency	Cost of equipment	What type of coating
Padding	5-20	Very little	Medium-difficult	Average	Medium	Low	Stucoo, preliminery, base coating, middle coating
Roller coating	2-40	Little	Medium-difficult	Very good	High	High	Preliminery, base coating, middle coating, top coating
Spraying	1-8	High	Easy-medium	Good	High	Medium	Preliminary, base coating, middle coating, top coating, finishing of special effect
Curtain coating	10~50	Medium	Medium-difficult	Good	High	High	Preliminary, base coating, top coating (patent leather)
HVLP	8-20	Medium	Medium	Good	Medium	Medium	Base coating, middle coating, top coating
Airless spraying	12-45	Medium	Medium	Good	Medium	Medium-high	Base coating, middle coating, top coating (patent leather)

Different finishing operations can be chosen and applied in the finishing process according to the state of the crust and the requirement of the resultant leather.

Compared with the spraying and rolling operation, the rolling operation is simple and can exactly control the coating weight so as to keep the stable quality of the finished leather. During the rolling operation, the compression between the roller and the crust can increase the adhesion between the finishing layer and the crust. Simultaneously it can save the finishing materials in the range of 30% and 40% so as to avoid the environmental pollution originated from spraying operation. Therefore in order to reduce the pollution in the finishing operation, rolling operation should be preferentially chosen and applied in the finishing process.

Questions

(1) Analyze the advantages and defects of the traditional methods of preservation of hides and skins and related clean technologies.

(2) Describe and analyze the defects in soaking, degreasing process of hides and skins and related clean technologies.

(3) Describe and analyze the defects in unhairing and liming process and relevant clean technologies.

(4) Describe and analyze the defects in deliming process and relevant clean technologies.

(5) Describe and analyze the defects in pickling process and relevant clean technologies.

(6) Describe and analyze the defects in chrome tanning process and relevant clean technologies.

(7) Describe the clean technologies in the chrome-free tanning process of pelt.

(8) Describe the clean technologies in the retanning, dyeing and fat-liquoring process of leather.

(9) Describe the clean technologies in the finishing process of leather.

Chapter 5 Resource utilization of tannery by-products

5.1 Introduction

5-1

The leather industry is a by-product industry and, as such, significantly reduces the environmental impact of the meat industry. Nevertheless, it is salutary to note that only about 20% of the original raw stock can in practice be converted into saleable leather, the remainder ends up as either wastes or by-products.

Limed splits produced as a by-product of beamhouse processing are utilized as a primary raw material by the collagen film and gelatin manufacturing industries. However, tanneries supplying splits for this purpose have process constraints imposed upon them: beamhouse processing must not involve enzymes, amines, or other substances not permitted by food regulations such as those implemented by the US Food and Drug Administration (FDA).

5.2 Collagen process method

Gelatine production involves an extended lime hydrolysis to break down the collagen into gelatins, which can then be extracted by means: of a series of hot water batches of increasing temperature at controlled pH. It is important to distinguish the gelatins breakdown of the collagen triple helix from the main chain scission where the gel strength would be severely affected.

The gelatin is subjected to purification, demineralization, concentration and sterilization prior to final drying. The major uses of collagen and gelatins are summarized below.

(1) Protective and aesthetic coating for meat and poultry.

(2) Sausage casing.

(3) Confectionery.

(4) Beer, wine and fruit juice clarification.

(5) Wound protection ("synthetic skin").

(6) Microsurgery.

(7) Dialysis.

(8) Sustained release drugs.

(9) Collagen injections for facial wrinkles, etc.

(10) Beauty products and toiletries.

5.3 Biogas

5.3.1 Biogas from hair treatment

A tannery processing 10,000 hides per week could be expected to produce 30 to 40 tonnes of hair per week if using a hair-saving process. Thus, with the increase in popularity of hair-saving processes, the exploitation of recycling technologies for separated intact hair is now becoming more important.

Composting to produce "biogas" (methane) is directly available to the tanner. The hair can be added to farm manure, flashings and effluent treatment sludge in a biogas reactor, where anaerobic activity produces methane gas and a fertilizer of high nutrient content.

In Denmark, a new outlet has proven practical for the fleshings and recovered hair from local tanneries operating hair save technologies. Initially, the plant was operated by tanker deliveries of pig manure and cow manure as a slurry of about 6% solids and an average temperature of 12℃. It was, however, found advantageous to add richer sources of energy in the form of stomach contents from two slaughterhouses and the solid wastes from the tanneries. With the fleshings macerated to a maximum size of 50 mm there are no problems of settling out in the holding tanks or poor degradation. The hair was found to be solubilized into solution in the sterilizing process.

5.3.2 Biogas process method

The manure, gut contents, fleshings and recovered hair are blended and circulated in a retaining tank, awaiting sterilization by heating to 55℃ in a large capacity tank to destroy pathogens, viruses and weed seeds. This is then followed by cooling to 38℃, the most suitable temperature for biological conversion. Problems with sulfides are familiar to tanners, with the solution often involving oxidation with manganese sulfate as a catalyst. Scrubbing hydrogen sulfide from the biogas could have solved the problem at this plant, but a biological approach was adopted instead.

As the biogas evolves, about 5% atmospheric air is added (with no possibility of combustion) and the mixture passes into a tower fitted with an inert plastic lance. Condensate from the gas collects on the lattice and some of the screened fertilizer is added as a nutrient. Aerobic bacteria living within the lattice structure convert the hydrogen sulfide into sulfur which is washed from the lattice by condensing moisture. It is removed daily by pumping into the final holding tank for return to the land as a vital component for plant growth.

Anaerobic digestion and the attendant production of methane require a holding time of 20 days in two sealed tanks. The plant operation is managed by removing 5% of the degraded sludge daily with equal volume being replaced from the sterilizing tank. The bio-degraded sludge is then cooled to 25℃ and stored in holding tanks. The conversion of organic solids into methane is high but any remaining insolubles are screened and pressed to produce a fibrous compost for horticultural/garden use. The liquid component is returned in the empty sludge tanker to the farmers as a fertilizer for agricultural use.

The gas is collected and regulated in a gasholder, with the energy used in two different ways.

(1) Part of the gas is used in running a Caterpillar generator that supplies 11,400 kWh per day of electricity directly to the national grid.

(2) Another 12,000 kWh of energy is converted in a natural gas heating plant, adapted for burning biogas, into water at 96℃ for supply to a heat plant run by the city. The water cycles through heat exchangers at a larger heat station and returns to the biogas plant, with the city selling hot water for heating greenhouse and domestic uses.

5.4 Glue and gelatin

5.4.1 Production of glue and gelatin

Gelatin today is manufactured from three main raw material sources: pigskin, ossein (which is decalcified, decreased and crushed raw bone), and bovine hide. Pigskin, as a raw material, is used extensively in the USA, and increasingly so in Europe. Since it is a much younger form of collagen (nine months is the average lifespan of a pig these days) only a short, mild acid (5%) soak is all that is required as pretreatment prior to gelatine extraction. Much grease is also extracted and this is sold to offset the higher cost of the raw material which, because it is fresh from the slaughterhouse, needs to be stored deep frozen until used. It is found that approximately 7-8 tonnes of pigskin (40% collagen) is required to yield one tonne of gelatine (90% protein).

Ossein has to some extent been replaced by pigskin by Continental producers because it is expensive to produce. Bones have to be collected (from as far away as India and Pakistan), cleaned of blood and meat, crushed, demineralized with dilute acid and degreased. This process takes a few days to perform, after which there is a lime soak of about 100 days, which ties up large amounts of raw materials, space and money.

Limed bovine hide is relatively plentiful, thanks to our leather industry, and is relatively cheap as a raw material compared with ossein. To process hide into gelatine, an alkali pretreatment is usually performed. Either a lime soak, similar to ossein pre-treatment, or a caustic soda-based process is used.

The manufacture of glue and gelatine from hide trimmings and chrome leather waste has for many years been a principal disposal route for these kinds of tannery by-products; however, there are no restrictions on the physical characteristics of the chrome leather waste.

5.4.2 Uses of gelatin

Gelatine is widely used in industries such as pharmaceuticals, food and wines, confectionery, photographic films, etc.

Gelatin is sold throughout the world on its main property, its Bloom value. This is a measure of its strength and is named after an American, Iscar T. Bloom, who invented a machine based on the weight of lead shot required to press a half-inch diameter plunger into the surface of a gelatin set under standard times and temperature conditions for 16-18 hours at 10℃. Modern equipment has replaced the lead shot Bloom machine, the latest device being the British Food Manufacturers Research Association' s texture analyzer, controlled by microprocessors. Bloom values for gelatins vary in practice but are generally classified into low Bloom (150 and below), medium Bloom (150-220) and high Bloom (over 220 up to about 300). There is a large and varied market available for the different qualities of gelatin. The economics depend on the quality of the product, with more being paid for pharmaceutical grade gelatin and less for photographic. This must be balanced against the cost of purification and waste disposal costs, the latter being high for gelatins from chrome shavings due to the chrome content.

5.4.3 Uses of glue

The major part of animal glue production goes into remoistenable gummed tapes for use on envelopes etc. Low grade glues are used in the making of paper cartons and to some furniture manufacturers animal glues are still a preferred product. There has been a large displacement of animal glue from the market place due to the comprehensive range of inexpensive synthetic resin and rubber adhesives that have been developed and continue to appear at an ever-increasing rate.

5.4.4 Glue process method

After the raw bovine material is delivered, it is cut into small pieces using large mechanical cutters and pumped into concrete soaking pits. It is subjected to a highly developed alkaline pretreatment process based on caustic soda, during which many of the covalent cross-links in the collagen triple helix are broken, leaving a product soft, swollen and extensively hydrated. The alkali remaining in the skins is removed by water washers. The final neutralization is performed with hydrochloric acid, after which the skins are washed back.

Washing is performed in large hide processors. Clean pieces and trimmings are pumped to specially designed extraction vessels, where the gelatin is extracted from the skins into hot water to a concentration of 5%, at increasing temperatures in steps of 5-10℃. This allows three of four separate extractions. The higher temperatures produce gelatin of lower gel strength and viscosity and increased color which commands lower product prices.

Some producers of pharmaceutical grades also use carbon filters to remove traces of biocides.

After primary filtration, the gelatin is improved to a sparkling clarity using a plate and frame filter fitted with filter card sheets. The liquor is subjected to deionization by ion exchange resins. Following deionization, the liquor is then evaporated to 30%-40% total solids using a double effect system which evaporates under vacuum, so keeping heat damage to a minimum. This also has the added effect of sucking out any volatiles, so giving gelatin of low smell and bland taste. This is in conjunction with a sterilizing system based on live stream injection which flashes the temperature of the gelatine up to 135℃ for 4-5 seconds.

The gelatine liquor is then chilled using a rotatory, scraped surface heat exchanger, which makes the gelatine solution set or gel. At the same time, a positive pump extrudes the set gelatine through a perforated plate, producing long spaghetti-like noodles with a high surface area, because of the fractured nature of the noodle exterior. The noodle shang down and break under their own weight, dropping onto a rubber-surfaced conveyor which oscillates from side to side, laying the gelatin onto a wire mesh belt. This belt feeds the noodles into an automatic drying tunnel which can dry gelatin at 30% total solids content to 90% total solids content in 2 hours. All being well, after two hours, the dry gelatin is pushed off the bed and broken up for storage and testing. Samples of every batch made are tested in the laboratory and customer blends are created from these results. The specifications are confirmed before batches leave the premises.

5.4.5 Gelatin from chrome shavings

The process for extracting gelatin from waste chrome shavings is simply one of raising the pH and washing with hot water to mobilize this protein. The gelatin is then purified by passing through an ion exchange column and concentrated. After cooling the gelatin may be extruded and dried.

Green or limed fleshings can be digested to yield a byproduct known as tallow subjected to one of the following treatments.

(1) Steam hydrolysis (the Lamatic process).

(2) Alkaline hydrolysis.

(3) Enzymatic digestion.

5.4.6 Case for treating fleshings by BLC

BLC first investigated the potential of enzymes for the rendering of fleshings in 1990, demonstrating the effectiveness and commercial viability on an industrial scale in 1991. Due to its success, the enzymatic treatment technology has now been successfully adopted by two tanneries in England, whilst less refined digestion techniques have been adopted with limited success in other countries.

The recovery of the tallow from green and limed fleshings reduces the volume of solid waste from these processes, as the water that is the major part of the fleshings (approximately 60%) is separated from the grease and protein components. As the tallow is extracted, this means that there is less fat in the effluent. Also, the resultant wastewater is alkaline in nature, and may be easily treated with the other waste streams in the tannery's effluent plant.

The use of enzymes ensures that the minimum amount of chemicals is used in the process. As these enzymes by their nature are proteins, they are easily denatured and biologically broken down.

The disadvantages of the process are three-fold. The first is that tallow recovery generates wastewater with a very high chemical oxygen demand. For the digestion stage, CODs are in the region of 80,000 ppm. The acid cracking of the grease emulsion gives an aqueous phase with a COD of approximately 30,000 ppm. The use of coagulants and flocculants has significantly reduced the COD content of the water discharge as well as improving settlement of grease and solids, but this adds extra cost to the process.

The second disadvantage is that of odor emission. During the acid cracking of the grease emulsion, the pH is reduced to 2. 0. For limed fleshings, any sulfide present would be converted to hydrogen sulfide giving a distinctive "rotten egg" odor. The addition of a powerful oxidizing agent, such as hydrogen peroxide, would reduce these odors and bleach the tallow, making an improved product. The use of concentrated acids and powerful oxidizing agents would need to be strictly controlled as mixing the two would have serious consequences and so qualifies as the third disadvantage.

Cost savings are gained from the reduction in transporting fleshings to landfills and the charge from the tips. These are constantly increasing in price, and the number of landfill sites that will accept fleshings is on the decline. As approximately 60% of limed fleshings are water the tallow recovery process separates this fraction for disposal through the wastewater treatment plant. A further logo of the fleshings is sold as tallow. This potential saving will quickly pay for the installation of equipment required for tallow extraction.

The BLC process is essentially a two-stage process. Once the fleshings have been chopped to increase their surface area, they are digested in a reactor by an enzyme (Savinase) at 55℃ for 30-60 minutes. Agitation of the mixture is then stopped, which allows an emulsion of tallow in water to separate as a well-defined upper layer.

For the second stage, the lower, sludge/aqueous layer is drained off, and the emulsion is transferred to a second reactor. Here, the emulsion is treated with hydrogen peroxide and acid (sulphuric or hydrochloric), and "cracked" by steam injection to bring the emulsion to boiling point. This liberates the tallow (typically containing 10% free fatty acids but only 0. 5% water), which is separated from the underlying aqueous phase. The efficiency of the system is high, with tallow recovery being in the order of 90%.

5. 5 Tallow recovery by mechanical means

A different approach to extract tallow without creating effluent disposal problems and reduce the volume of fleshings has been transferred from the operation of screw presses in the extraction of edible oils from seeds.

The basic design and operation of the plant is very simple. Designed to treat both green and limed fleshings, delivery is either directly from the fleshing machine or by pump into the feed hopper. The press comprises twin intermeshing screws that rotate in opposite directions and these are fitted within a heavy duty perforated metal cage. The fleshings are transported through the machine as the screws rotate by the pitch of flights, which decreases towards the discharge end. This causes the increase of pressure on the flushings, causing liquid (containing up to 20% tallow) to be expelled through the cage.

Several options are open for the treatment of the extracted liquid, including: three-phase decanting, flotation, gravity separation.

All ensure that none of the tallow is allowed into the effluent treatment plant. It has been noted that tallow can be extracted in this way at temperatures as low as 15℃, although causes some problems with blockages.

5.6 Chrome contained shavings

Environmental regulations demand that the chromium content in industrial wastewater should be less than limits. This means that chromium in tannery wastewater needs to be necessarily treated prior to discharge. Chrome-containing sludge is classified among hazardous wastes. For the disposal of chrome sludge, safe locations are identified by municipal authorities. In the event, such locations are not specified; the individual tanneries have to provide adequate space within the unit. Already this has been causing serious space problems.

5.6.1 Options for disposal

For the utilization of chrome sludge, three possible ways can be considered.

(1) Utilizing the combustion energy of the organic fraction by adding a mixture of bricks and tiles.

(2) Utilizing the chrome content for the production of high value ceramics, e.g. refractory materials after pretreatment of the sludge.

(3) Utilizing the chrome content for the production of pigments based on chromium after pretreatment and enrichment of the sludge.

Option (1) should provide a safe method of depositing chromium in an industrial product without necessarily exploiting the economic value of chromium. Options (2) and (3) on the other hand envisage the exploitation of the economic values of chromium in either adding special heat transfer properties or special color to the surface coatings.

As a result of literature studies and some preliminary trials concerning the characterization of typical industrial chrome sludges, options (2) and (3) are considered unrealistic and unviable under the present circumstances.

5.6.2 Benefits

From an environmental engineering perspective, safe sludge inclusion within brid achieves the following benefits.

There is the obvious result of positively incorporating waste sludge residues within useful products (e.g. brick), thereby contributing to a reduction in the sheer volume of the unwanted, unwholesome commodity.

Pathogens found within the sludge will assuredly be destroyed during firing.

Organics contained within the sludge could be fullyoxidized given the residency time (e.g. in the order of days, rather than seconds within an incinerator), firing temperature (up to 1,050℃), and gas recycling systems provided within a brick kiln.

On the other hand, the brick manufacturer may also realize a number of operational benefits in conjunction with the use of sludge, which are as follows.

(1) The previously mentioned combustion of sludge organic yields heat release and, consequently, reduces the energy required for firing of the brick.

(2) The incorporation of wet sludge within a brick manufacturing process will obviate the need for adding moisture to the clay, which is typically necessary to develop its plasticity prior to extrusion.

(3) The combustion of sludge materials may introduce small voids within the body of the brick; these voids may improve both the brick's response to freeze-thaw expansion and its bonding adherence to mortar.

5.7 Sludge

5.7.1 Sludge processing

Tannery sludge as obtained is difficult to further grind to finer particles due to the great toughness of the material. However, when silica-containing clay is used as a grinding medium, the sludge can be sized down in a practicable way to fine particles. Brick manufacture in this country is generally by two methods.

(1) Wire-cut bricks or extruded bricks.

(2) Country-made bricks.

The Bureau of Indian Standards has fixed certain standards for wire-cut bricks made through extrusion and firing in organized factories. Usually, such bricks have cold crushing strength of 75-100 kg/cm^2, water absorption of 14%-16%, and bulk density of 1.8-1.85 kg/dm^3. This category takes care of more than 75% of the total requirement of the bricks in the country.

The clay or mixture of clays are made into a paste having 20%-25% moisture and hand modeled in wooded moulds or shaped by a manually operated process to 'green bricks'. There are different sizes and shapes followed as a practice in different parts of the country. The bricks are stacked one over the other with intermediate gaps and then the firing is done in an up draught fashion in some spaces. Wood logs are fed into the fire places. Coal dust is spread between the brick layers. Irrespective of the method of firing, 15%-30% of the bricks are generally unfired or less fired, which are refired in subsequent firings.

Country-made bricks are also covered by the Bureau of Indian Standards. However, this industry is widely distributed in tiny workplaces and hence the specifications are seldom followed. These bricks are low strength (15-30 kg/cm^2) with water absorption (15%) and density of 1.75-1.85 kg/cm^2.

Sludge-amended brick production appears to offer several different benefits relative to constructive sludge management. In view of the magnitude of Indian brick production (approx. 40,000-45,000 million pieces in 1992), processing yearly 90-1,000 million tonnes of clay, the application of 130,000 tonnes of chrome sludge must be considered feasible.

5.7.2 Health and safety considerations

Recent experiments on sludge/clay mixtures including large scale production experiments lead to the conclusion that under the usual oxidative conditions up to 90% of the Cr (Ⅲ) added can be oxidized to Cr (Ⅵ). Another important observation concerning environment and worker protection was the production of "foul gases".

From reports and publications, we may conclude that the uncontrollable utilization of chromium-containing tannery sludges in brick manufacturing is not without risks for workers and the environment.

In any case, contact of Oman tissue with Cr (Ⅵ) must be avoided, and contact with Cr (Ⅲ) must be minimized to an acceptable level.

The emission of gases from the die kiln and the stack during the early stage of firing of the sludge-clay mixtures should be controlled adequately.

5.7.3 Conclusions on methods processing sludge

The products manufactured from a sludge/clay mixture are lighter and more porous than the products made from clay alone, by controlling the addition and firing conditions it is possible to attain strengths above the Indian Standard Specifications.

The bricks fired under oxidizing conditions at the optimum temperature of 900℃ gave a brick red color, bulk density of 1,600-1,700 kg/cm^3 and water absorption of 18%-25%.

To obtain acceptable mechanical properties the temperature of firing should be around 900℃ depending on the clay mass.

The bricks thus produced under the above process have a net energy saving of about 20% during firing because of the burning of the organics in the tannery sludge. In fact, this behavior is comparable with similar additives such as coconut pith and sawdust.

It is possible to make sludge/clay mixtures either by hand forming, by extrusion or by semi-dry pressing.

The tannery sludge can be easily homogenized with plastic clays up to about 15% by weight, using layer-stacking of the raw materials, followed by pan milling. The sand present in the clay acts as a grinding medium.

The extruding bricks made out of the sludge/clay mixture have higher shrinkage up to about 50% compared to usual clays.

The production of chrome-sludge-containing bricks requires slower firing schedules up to 700℃ in order to ensure the complete burning of organics.

Pre-conversion of Cr (Ⅲ) to chrome (Ⅲ) oxide under reducing conditions at 450℃ does not hamper Cr (Ⅵ) formation during subsequent oxidative firing.

Oxidative firing at elevated carbon dioxide pressure does not hamper alkalinity increase and Cr

(Ⅵ) formation.

A reductive atmosphere during the cooling schedule has a profound influence on the control of conversion of Cr(Ⅲ) to Cr(Ⅵ). Studies by TNO confirmed the reduction atmosphere during cooling the fired bricks to less than 100℃ is most important to convert the Cr(Ⅵ) formed into Cr(Ⅲ) adequately and keep it in this form.

The leach ability of Cr(Ⅵ) in the bricks containing sludge-clay mixtures is high when fired in an oxidizing atmosphere and for long periods, while the leach ability is strongly reduced during firing in a reducing atmosphere. The leach ability comes below the standards when fired in a fully reducing atmosphere. This, however, calls for controlled and practicable firing conditions.

The study has ultimately shown that it is possible to incorporate tannery sludge safely in bricks. Under practicable technical processing conditions, bricks were made containing negligible amounts of Cr(Ⅵ) in the product.

In this way, a safe and useful disposal of the environmentally hazardous solid waste of the leather industry has been demonstrated to be practicable under Indian working conditions.

Questions

(1) Give your descriptions about the usage of gelatin.

(2) Give your descriptions about the technology of producing gelatin from limed splits.

(3) Give your descriptions about the technology of Gelatin extraction from chrome shavings.

(4) Give your descriptions about the utilization of hair from tannery.

Chapter 6 Environmental protection and regulations

6.1 Environmental protection related to leather processing

6.1.1 Water pollution in leather production

6-1

The government which is responsible for the protection of environment & nature and reactor safety, has classified the pollutants according to their potential hazard in a catalogue. Water pollutants are solid, liquid and gaseous substances which are capable of changing the physical, chemical or biological conditions of water to a lasting extent. For leather manufacture, water pollution is the main issue to be concerned.

The assessment of the water-polluting potential is based on the specific properties of the substances, which are as follows.

(1) Acute toxicity on mammals.

(2) Aquatic toxicity on fishes, algae and bacteria.

(3) Biological degradability (hydrolysis, photolysis, oxidation, etc.).

(4) Soil mobility.

(5) Carcinogenic effect.

(6) Mutagenic effect.

(7) Teratogenic effects.

6.1.2 Environmental protection for leather production

For leather manufacture, environmental protection is firstly started with the selection of environmentally friendly products and processes, and continues with appropriate treatment of waste water and airborne emissions. According to the technology roadmap for water-saving and emission-reduction in leather industry of China, the selection of environmentally friendly products and clean processes in leather production are suggested in Table 6.1.

Table 6.1 Environmentally friendly products and processes in leather production

Operations	Clean technologies
Soaking	Fresh raw hides free of salt (green hides) Biodegradable surfactants

Continued Table

Operations	Clean technologies
Liming	Low sulfide and sulfide-free liming processes Hair saving processes Recycling of spent liquor
Deliming	Low ammonium and ammonium-free deliming processes
Bating	Low ammonium and ammonium-free deliming processes
Pickling	Low salt and salt-free pickling processes
Tannage	Recycling of residual chromium in spent tanning liquor from wet blue production improved chrome exhaustion and fixation Pretanning and shaving of the pretanned pelts Alternative tanning techniques without chromium
Retanage	Retanning agents with high exhaustion Polymeric tanning agents Low-phenol and low-formaldehyde syntans Low-salt liquid retanning agents Low-formaldehyde resin tanning agents Aldehyde tanning agents
Dyeing	Deducted powder dyes Liquid dyes Fixing agents Dyeing auxiliaries
Fatliquoring	AOX-free fatliquors Polymeric fatliquors with high exhaustion
Water repellent	Water repellents that do not need to be fixed with metal salt treatment
Finishing	Aqueous finishing systems Pigment preparations free of heavy metals

6.1.3 Regulations concerning substances contained in effluents

Regulations governing the quality of effluents discharged from tanneries, differ from one country to another. The restrictions are more or less stringent, depending on the local conditions and whether the effluents are discharged into a main drain directly or indirectly, i. e. first into a public waste water or central treating plant. Samples of the effluents for testing are taken at the point at which the effluents are discharged into the public sewer, or from the main drain. To better meet the requirements of water regulations, the contests in Table 6.2 offer methods for adjusting the most important effluent parameters.

In China, installations for storing, filling and treating water-polluting substances and installations using water-pollutants in industry and public facilities, have to be so constructed that waters are neither polluted nor lastingly changed in their properties, thus to meet the requirements of the mandatory Standard GB 30486, published in 2013. The limited values for each effluent parameter based on the regulation are listed in Table 6.3.

Table 6.2 Regulations requirements and water treatments

Toxic substances	Water treatment technologies
COD/BOD	Cut down oxygen demand by flocculation, sedimentation and biological degradation. Omit use of oxygen-demanding substances
Ammonia/nitrogen	Use hair saving liming processes and nitrogen-free products (particularly in deliming)
Aluminium	Flocculate out of residual and wash liquors
Chromium (Ⅲ)-compounds	Flocculate chromium salts out of residual and rinsing liquors, recalculate residual liquors, fix chromium salts in the leather, and use chrome tanning process with extensive exhaustion of residual liquors
Phenol	Use products with the lowest content of free phenol
Sulfide	Use sulfide-free liming chemicals, use catalytic safety, environmental protection. Oxidizing agents (manganese sulfate, manganese chloride), recirculate residual liming liquors. Recover sulfides by acidifying residual liquors and running all of the hydrogen sulfide formed into caustic soda
Chloride ions	Use products with lower Cl^- content in Soaking or in the pickling in place of NaCl
AOX	Use products that do not contain organic halogen compounds

Table 6.3 Discharge standard of water pollutants for leather and fur-making industry (GB 30486-2013)

Effluent parameters	Existing tanneries[a]		Newly-built tanneries[a]	
	Drain direct	Drain indirect	Drain direct	Drain indirect
pH value	6-9	6-9	6-9	6-9
Color degree	50	100	30	100
Suspended solids	80	120	50	120
BOD_5	40	80	30	80
COD_{Cr}	150	300	100	300
Oils & fats	15	30	10	30
Sulfide substances	1	1	0.5	1.0
Ammonia nitrogen	35	70	25	70
Total nitrogen	70	140	40	140
Total phosphorus	2	4	1	4
Chloride ion	3000	4000	3000	4000
Total chromium	1.5	1.5	1.5	1.5
Hexavalent chromium	0.2	0.2	0.1	0.1
Water consumption	65 m^3 per ton of raw hide		55 m^3 per ton of raw hide	

a) The units are "mg/L", except for pH value and color degree.

6.1.4 Air pollution

6.1.4.1 Airborne emissions

Airborne emissions are in many countries less strictly regulated than discharges in waste water. All countries have their own classification and limitations of hazardous substances that are released into air. No specific limits exist for the leather industry. In China, airborne emissions are covered by the law of the People's Republic of China on the Prevention and Control of Atmospheric Pollution. The emissions to air in the leather industry, and the related avoidance are summarized in Table 6.4.

Table 6.4 Emissions to air in the leather industry

Process	Substance	Avoidance
Wet end	Dust	Dedusted products Liquid products
Deliming/Bating	H_2S	See below
Pretanning/Tanning/ Retanning	Glutaraldehyde	Automatic dosage systems Modified Glutaraldehyde with lower vapor pressure
Finishing	VOC	Low VOC-systems Water-based systems

6.1.4.2 Hydrogen sulfide-H_2S

The use of sodium sulfide, sodium hydro-sulfide and organic sulfides for the unhairing process is likely to cause the development of H_2S in tanneries.

H_2S is strongly poisonous. It has an irritating effect on the mucous membranes and paralyzes cell respiration and thus damages the nerves. Poisoning by H_2S results in inflammation of the eyes, bronchial catarrh and inflammation of the lungs. Higher concentrations cause cramps, unconsciousness and eventually death due to respiratory paralysis.

Even at concentrations as low as 700 ppm. H_2S in the inhaled air, acute lethal poisoning may occur after a short time. In addition, H_2S gas forms explosive mixtures with the air (explosion limits: 4.3%-45.5% by volume, ignition temperature 270℃). Therefore, it is absolutely necessary to avoid ignition sources.

Some measures have been applied to reduce the likelihood of gas contacting H_2S as follows.

(1) Use low sulfide or sulfide-free liming methods.

(2) Wash limed pelts several times in long floats (200%) to reduce sulfide content pelts.

(3) Incorporate sodium bisulfite in delimiting to oxidize H_2S.

(4) Wash thoroughly after deliming and bating.

(5) Installation of ventilation and exhaust devices on drums and in workrooms.

(6) Wear a protective mask when working at the drum.

(7) Measure sodium sulfide concentration at the workplace and the door of drums with the test tube for sodium sulfide.

6.2 Critical substances in leather

Besides environmental requirements, restricted substances in consumer products have also been an important issue for both manufacturers and consumers. Their use is limited for a number of reasons including consumer safety and worker safety. Many brand name manufacturers require certificates of compliance about these critical substances from their supplier.

Regulations and standards restrict the use of toxic chemicals in consumer products made of leather. Leather industry has already taken action by replacing the restricted substances or assuming the limits imposed by these restrictions. These regulations and Standards are listed as follows.

(1) GB 20400 by China.

(2) European REACH regulations.

(3) US Consumer Product Safety Improvement Act, CPSIA.

(4) American Apparel Footwear Association, AAFA-RSL.

(5) Zero Discharge of Hazardous Chemicals, ZDHC-MRSL.

(6) Leather by OEKO-TEX 100.

(7) Blue-Angel DE-UZ 148 by German.

(8) ISO 20137.

However, lists of restricted chemicals by regulations or standards, contain many substances that are not relevant to the leather industry. Among these regulations and Standards, only ISO 20137 reveals a relatively objective and full description of critical chemical residues in leather. This document gives an overview of these internationally accepted chemicals contained in leather, as listed in Table 6.5. For those chemical substances that are not mentioned in Table 6.5, they should not be concerned with leather, thus avoiding unnecessary costs.

Table 6.5 Overview of critical chemicals contained in leather, their possible uses and restriction

No.	Substance	Possible uses	Mandatory restriction or legal obligation
1	Alkylphenol, AP alkylphenol ethoxylates, APEO	Soaking, degreasing, and finishing. However, AP would not be used in leather industry	REACH EU regulation No. 1907/2006, Annex XVII-entry 46, in the treatment of leather
2	Aromatic amines released from AZO dyes	Colorants	REACH EU regulation No. 1907/2006, Annex XVII-entry 43, GB 20400
3	Cadmium	Only for coated leather	REACH EU regulation No. 1907/2006, Annex XVII-entry 23, SVHC
4	Chlorinated paraffin, SCCP	Oil tanning and fatliquor formulations	REACH EU regulation No. 1907/2006, SVHC candidate substance. Persistent organic pollutant (POP)

Continued Table 6. 5

No.	Substance	Possible uses	Mandatory restriction or legal obligation
5	Chlorophenols -Penta-chlorophenol, PCP -Tetra-chlorophenol, TeCP -Tri-chlorophenol, TCP	Biocides	REACH EU regulation No. 1907/2006, Annex XⅦ-entry 22 Biocide EU Regulation No. 528/2012
6	Dimethyl formamide, DMFa	Only for coated leather. Solvent for PU	REACH EU regulation No. 1907/2006, SVHC candidate substance
7	Dimethyl fumarate, DMFu	Only used in the package of upholstery and shoes to protect them from mould.	REACH EU regulation No. 1907/2006, Annex XⅦ-entry 61
8	N-methyl pyrrolidone, NMP	Only for coated leather	REACH EU regulation No. 1907/2006, SVHC candidate substance
9	Phthalates	Only for coated leather. Plasticizers/softening agents, used in PVC and PU finish coat formulations and fatliquors	REACH EU regulation No. 1907/2006, SVHC candidate substance
10	Polyaromatic hydrocarbons, PAHs	Only for coated leather	REACH EU regulation No. 1907/2006, Annex XⅦ-entry 50
11	Bisphenol A	Retanning agent	REACH EU regulation No. 1907/2006, SVHC candidate substance
12	Chromium (Ⅵ), Cr (Ⅵ)	Not used for leather tanning. Traces of Cr (Ⅵ) oxidation state can develop if oxidative conditions are allowed to occur	REACH EU regulation No. 1907/2006, Annex XⅦ-entry 47, SVHC candidate substance
13	Flame retardants -Polybrominated diphenyl ether	Finishing agent	REACH EU regulation No. 1907/2006, Annex XⅦ-entry 45 POP EU regulation No. 850/2004
14	Heavy metals -Arsenic, As -Barium, Ba -Mercury, Hg	Pigment	EU directive 2009/48/EC, on toy safety
15	Heavy metals -Lead, Pb	Pigment	REACH EU regulation No. 1907/2006, SVHC candidate substance. California proposition 65 List US Consumer Product Safety Improvement Act, CPSIA
16	Organotin compounds	Fungicides in certain auxiliaries. Catalysts for PU synthesis	REACH EU regulation No. 1907/2006, Annex XⅦ-entry 20 EU directive 2009/48/EC, on toy safety

Continued Table 6. 5

No.	Substance	Possible uses	Mandatory restriction or legal obligation
17	Perfluoro octyl sulfonic acid, PFOS	Soil, oil and water-resistant products	POP EU regulation No. 850/2004

Questions

(1) Why should the effluents from tanneries be treated?

(2) Please give a description about clean processes in leather production according to the operations in details.

(3) Please give a description about the discharge standard of water pollutants for leather making industry according to GB 30486-2013 in China.

(4) Can hydrogen sulfide be effectively controlled or reduced during leather processing? Please give a description.

(5) Please give your comments in details about the critical chemicals residuals in leather according to current standard ISO 20137.

Reference

[1]COVINGTON A D. Tanning Chemistry:The Science of Leather[M]. RSC Publishing,2011.

[2]ZHANG SHUHAU,et al. History of China Leather Industry[M]. Beijing:China Social Sciences Press,2016.

[3]SUN DANHONG,HUNAG YUZHEN,GUO MENGNENG,et al. Color Histological Atlas of the Chinese Cowhide[M]. Chengdu: Sichuan science and technology press,2005.

[4]LI ZHIQIANG,LIAO LONGLI. The chemistry and histology of Animal skins[M]. Beijing:China Light Industry Press,2010.

[5]廖隆理. 制革化学与工艺学·上册[M]. 北京:科学出版社,2005.

[6]单志华. 制革化学与工艺学·下册[M]. 2版. 北京:科学出版社,2017.

[7]CHEN WUYONG,LI GUOYING. Tanning Chemistry[M]. Beijing:China Light Industry Press, 2018.

[8]MICHEL A. Alternative Technologies for Raw Hide and Skins Preservation[J]. Leather Ware, 1997,12(2):20.

[9]于淑贤. 现代生皮保藏技术文献综述[J]. 中国皮革,1999,28(17):23-25.

[10]BAILEY D G. The Preservation of Hides and Skins [J]. Journal of the American Leather Chemists Association,2003,98:308-318.

[11]HUGHES I R. Temporary Preservation of Hides Using Boric Acid [J]. Journal of the Society of Leather Technologists and Chemists,1974,58:100-103.

[12]KANAGARAJ J,JOHN SUNDAR V,MURALIDHARAN C,et al. Alternatives to sodium chloride in prevention of skin protein degradation—a case study [J]. Journal of Cleaner Production, 2005,13(8):825-831.

[13]单志华. 食盐与清洁防腐技术[J]. 西部皮革,2008,30(12):28-33.

[14]MNEY C A,CHANDRABAUBU N K. 降低制革废水中的盐含量[C]//国际皮革科技会议论文选编(2004—2005). 中国皮革协会,2005:18-22.

[15]RUSSELL A E. LIRICURE - Powder Biocide Composition for Hide and Skin Preservation [J]. Journal of the Society of Leather Technologists and Chemists,1997,81:137.

[16]RUSSELL A E. The LIRICURE Low Salt Antiseptic Delivery System [J]. World Leather,1998, 11(5):43.

[17]CORDON T C,JONES H W,NAGHSKI,et al. Benzalkonium Chloride as a Preservative for Hide and Skin [J]. Journal of the American Leather Chemists Association,1964,59:317-326.

[18]VEDARAMAN N,SUNDAR V J,RANGASAMY T,et al. Bi-Additives Aided Skin Preservation—an Approach for Salinity Reduction[C]//The XXIX Congress of the IULTCS and the 103rd Annual

Convention of the ALCA. Washington D, USA, 2007.

[19] VENKATCHALAM P, SADULLA S, DURAISWAMY B. Further Experiments in Salt-Less Curing [J]. Leather Science, 1982, 29: 217-221.

[20] VANKAR P S, DWIVEDI A, SARASWAT R. Sodium sulphate as a curing agent to reduce saline chloride ions in the tannery effluent at Kanpur: A preliminary study on techno-economic feasibility [J]. Desalination, 2006, 201(1/2/3): 14-22.

[21] 彭必雨. 制革前处理助剂: Ⅱ 防腐剂和防霉剂 [J]. 皮革科学与工程, 1999, 9(3): 53-56, 34.

[22] DIDATO D T, STEELE S R, STOCKMAN G B, et al. Recent developments in the short-term preservation of cattle hides [J]. Journal of the American Leather Chemists Association, 2008, 103: 383-392.

[23] MUNZ K H. Silicates for raw hide curing [J]. Journal of the American Leather Chemists Association, 2007, 102: 16-21.

[24] MUNZ K H. Silicates for raw hide curing and in leather technology [C]//The XXIX Congress of the IULTCS and the 103rd Annual Convention of the ALCA. Washington DC, USA, 2007.

[25] KANAGARAJ J, CHANDRA BABU N K, SADULLA S, et al. Cleaner techniques for the preservation of raw goat skins [J]. Journal of Cleaner Production, 2001, 9(3): 261-268.

[26] KANAGARAJ J, CHANDRA BABU N K, SADULLA S, et al. A new approach to less salt preservation of raw skin hide [J]. Journal of the American Leather Chemists Association, 2000, 95: 368-374.

[27] BAILEY D, GOSSELIN J A. The preservation of animal hides and skins with potassium chloride: A kalium Canada, ltd. technical report [J]. Journal of the American Leather Chemists Association, 1996, 91: 317-333.

[28] ROSS G D, BAILEY D, DIMAIO G, et al. Electron beam irradiation preservation of cattle hides in a commercial-scale demonstration [J]. Journal of the American Leather Chemists Association, 2001, 96: 382-392.

[29] BAILEY D G. Evergreen Hides Market Ready [J]. Leather Manufacture, 1997, 115(6): 22.

[30] WATERS P J, STEPHENS L J, SURRIDGE C. Controlled drying of australian raw wool-skins for long-term Preservation [C]//. Proceedings of International Union of Leather Technologists and Chemists Societies Congress, London, 1997.

[31] BAILEY D G. Ecological concepts in raw hide conservation [J]. World Leather, 1995, 8(5): 43.

[32] BaAILEY D G. Future Tanning Progress Technologies Proceeding of United Nations Industrial Development Organization Workshop [Z]. 1999.

[33] SAUER O. Experience over two year's fresh hide processing [J]. Journal of the Society of Leather Technologists and Chemists, 1992, 76(2): 68-70.

[34] BAILEY D, HOPKINS W. Cattlehide Preservation with sodium sulfite and acetic acid [J]. Journal of the American Leather Chemists Association, 1977, 72: 334-339.

[35] SEHGAL P K, PREETHI V, RATHINASAMY V, et al. Azardirachta Indica: A Green Material for

curing of hides and skins in leather processing[J]. Journal of the American Leather Chemists Association,2006,101:266-273.

[36]KANAGARAJ J,SASTRY T P,ROSE C. Effective preservation of raw goat skins for the reduction of total dissolved solids [J]. Journal of Cleaner Production,2005,13(9):959-964.

[37]MITCHELL J W. Prevention of bacterial damage brine cured and fresh cattlehides[J]. Journal of the American Leather Chemists Association,1987,82:372-382.

[38]STOCKMAN G,DIDATO D,HURLOW E. Antibiotics in hide preservation and bacterial control [J]. Journal of the American Leather Chemists Association,2007,102:62-67.

[39]COVINGTON A D,SHI B. High stability organic tanning using plant polyphenols. Part 1 the interactions between vegetable tannins and aldehydic crosslinkers [J]. J. Soc. Leather. Technol. Chem. ,1998,82(2),64-71.

[40]HUGHES I R. Temporary preservation of Hides Using Boric Acid [J]. Journal of the Society of Leather Tecnologists and Chemists,1974,58:100-103.

[41]KANAGARAJ J,JOHN SUNDAR V,MURALIDHARAN C,et al. Alternatives to sodium chloride in prevention of skin protein degradation—a case study [J]. Journal of Cleaner Production, 2005,13(8):825-831.

[42]LUO JIANXUN,FENG YAN JUAN. Cleaner chrome tanning-technology of chrome-reduced tanning without salt, pickling and short procedure [J]. Journal of Society of Leather Technologists and Chemists,2019,103(6):289-295.

[43]LUO JIANXUN,MA HEWEI,FENG YANJUAN. Synthesis of an amphoteric polymer auxiliary agent and its application to chrome-free leather[J]. Journal of Society of Leather Technologists and Chemists,2018,102(6):298-303.

[44] LUO JIANXUN, FENG YANJUAN. Cleaner processing of bovine wet-white: Synthesis and application of a novel chrome-free tanning compound [J]. Journal of the Society of Leather Technologists and Chemists,2015,99:190-196.

[45] LUO JIANXUN, FENG YANJUAN. A novel eco-combination tannage of chrome-free leather with softness and high shrinkage temperature[J]. Journal of Society of Leather Technologists and Chemists,2022,106(3):99-105.

[46]Re-utilization of Biomass Resources:Preparation and Application of a Bio-polymer Retanning Agent Based on Cattle Hair Hydrolysate [J]. Journal of Society of Leather Technologists and Chemists, 2020,104(1):39-43.

[47] LUO JIANXUN, SHAN ZHIHUA. Wet-white leather processing: A new complex combination tannage[J]. Journal of Society of Leather Technologists and Chemists,2011,95(2):93-97.

[48] LUO JIANXUN, FENG YANJUAN, SHAN ZHIHUA. complex combination tannage with phosphonium compounds, vegetable tannins and aluminium tanning agent [J]. Journal of Society of Leather Technologists and Chemists,2011,95(4):215-220.

[49] SHRIVASTAVA H Y, NAIR B U. Chromium (Ⅲ) - mediated structural modification of glycoprotein: Impact of the ligand and the oxidants [J]. Biochemical and Biophysical Research

Communications,2001,285(4):915-920.

[50]FATHIMA N,RAO J,NAIR B. Chromium(Ⅵ)formation:Thermal studies on chrome salt and chrome tanned hide powder [J]. Journal of the American Leather Chemists Association,2001, 96:444-450.

[51] BALAMURUGAN K, VASANT C, RAJARAM R, et al. Hydroxopentaamminechromium(Ⅲ) promoted phosphorylation of bovine serum albumin:Its potential implications in understanding biotoxicity of chromium [J]. Biochimica et Biophysica Acta,1999,1427(3):357-366.

[52]彭必雨．钛鞣剂、鞣法及鞣制机理的研究 [D]．成都:四川大学,1998.

[53]郭文宇,单志华．一种膦盐鞣剂的开发以及应用前景 [J]．中国皮革,2004,33(3):1-4,48.

[54]范浩军,何强,彭必雨,等．纳米 SiO_2 鞣革方法和鞣性的研究 [J]．中国皮革,2004,33(21):37-38,42.

[55]鲍艳,杨宗邃,马建中．示差扫描量热法(DSC)研究乙烯基类聚合物/蒙脱土纳米复合鞣剂的鞣制机理 [J]．中国皮革,2006,35(19):14-17.

[56]HERNANDEZ J F,KALLENBERGER W E. Combination tannage with vegetable tanning and aluminium [J]. J. Amer. Leather Chem. Ass. ,1984,79(2),182-206.

[57]单志华．Re-MT 结合鞣法研究 [D]．成都:四川大学,1997.

[58]单志华,辛中印．无金属鞣制研究:有机结合鞣法 [J]．皮革科学与工程,2002,12(5):18-21.

[59]张廷有．鞣制化学 [M]．成都:四川大学出版社,2003:17-26.

[60]石碧,何先祺,张敦信,等．水解类植物鞣质性质及其与蛋白质反应的研究:Ⅳ植物鞣质与氨基酸的反应 [J]．皮革科学与工程,1994,4(1):18-21.

[61]何先祺,王远亮．植—铝结合鞣机理的研究[J]．中国皮革,1996,25(6):15-18.

[62]陈武勇,李国英．鞣制化学[M].2 版．北京:中国轻工业出版社,2005.

[63]COVINGTON A. Theory and mechanism of tanning:Present thinking and future implications for industry[J]. J. Amer. Leather Chem. Ass. ,2000,85(1):24-34.

[64]石碧,范浩军,何有节,等．有机鞣法生产高湿热稳定性轻革[J]．中国皮革,1996,25(6):3-9.

[65]D'AQUINO A,BARBANI N,D'ELIA G,et al. Combined organic tanning based on mimosa and Oxazolidine development of a semi-industrial scale process for high-quality bovine upper leather[J]. J. Soc. Leather. Technol. Chem. ,2004,88(2):47-55.

[66]陆忠兵,廖学品,孙丹红,等．单宁—醛—胶原的反应:对植醛结合鞣机理的再认识[J]．林产化学与工业,2004,24(1):7-11.

[67]罗建勋．多元复合结合鞣革稳定性的研究[D]．成都:四川大学,2006.

[68]吕绪庸．植物鞣剂简史及其取代铬鞣的技术建议[J]．西部皮革,2006,28(10):20-22.

[69]兰云军,刘建国,周建飞,等．无铬鞣技术规模化推广应用的可行性[J]．中国皮革,2008,37(13):44-46,54.

[70]蒋岚,吴瀚,史楷岐,等．THP 盐在皮革鞣制中的应用[J]．中国皮革,2007,36(23):54-57.

[71]TRAUBEL H,NOVOTNY F,REIFF H. Process for tanning leather:US5820634[P]. 1998-10-13.

[72]蒋岚,史楷岐,李颖,等．丙烯酸树脂与 THP 盐结合鞣制工艺的研究[J]．中国皮革,2006,

35(15):23-26.

[73] COVINGTON A, LAMPARD G, HANCOCK R A, et al. Studies on the origin of hydrothermal stability: A new theory of tanning[J]. Journal of the American Leather Chemists Association, 1998, 93: 107-120.

[74] FATHIMA N NISHAD, ARAVINDHAN R, RAO J RAGHAVE, et al. Tannic acid-pho sphonium combination: Aversatile chrome-free or ganictanning[J]. Journal of the American Leather Chemists Association, 2006, 101(5): 161-168.

[75] WINDUS W, H APPICH F W. A new tannage tetrakis (hydroxymethyl) phospho nium-resorcinol[J]. J. Am. Leather. Assoc., 1963, 58(6): 638-653.

[76] WINDUS W, FILACHIONE Edward M, HAPPIC F W. Combination tannage with tetrakis (hydroxymethyl) phosphonium and phenol: US, 3104151[P]. 1961-06-30.

[77] 四川大学,四川亭江新材料股份有限公司. 无盐不浸酸无铬鞣剂 TWT 鉴定资料[Z].

[78] 张铭让,陈武勇. 鞣制化学[M]. 北京:中国轻工业出版社,1999.

[79] 石碧,狄莹. 植物多酚[M]. 北京:科学出版社,2000.

[80] 张廷有. 鞣制化学[M]. 成都:四川大学出版社,2004.

[81] HERNANDEZ J F, KALLENBERGER W E. Combination tannage with vegetable tanning and aluminium[J]. J. Amer. Leather Chem. Ass., 1984, 79(2): 182-206.

[82] 单志华. 制革化学与工艺学[M]. 北京:中国轻工业出版社,1999.

[83] 杨德. 试验设计与分析[M]. 北京:化学工业出版社,1999.

[84] 成都科学技术大学西北轻工业学院. 制革化学及工艺学[M]. 北京:中国轻工业出版社,1982.

[85] IUP 35: Measurement of dry heat resistance of leather[J]. J. Soc. Leather. Technol. Chem., 1989, 73: 373.

[86] D AQUINO, A BARBANI, N D ELIA G, et al. Combined organic tanning based on mimosa and Oxazolidine development of a semi-industrial scale process for high-quality bovine upper leather[J]. J. Soc. Leather. Technol. Chem., 2004, 88(2): 47-55.

[87] 强西怀,毛李蓉,张辉,等. THPS 的改性与鞣性的相关性研究[J]. 中国皮革,2007,36(23):17-19,23.